The Code of the City

Urban and Industrial Environments

Series editor: Robert Gottlieb, Henry R. Luce Professor of Urban and
Environmental Policy, Occidental College

The Code of the City

Standards and the Hidden Language of Place Making

Eran Ben-Joseph

The MIT Press   Cambridge, Massachusetts   London, England

MIT Press books may be purchased at special quantity discounts for business or sales promotional use. For information, please e-mail special_sales@mitpress.mit.edu or write to Special Sales Department, The MIT Press, 5 Cambridge Center, Cambridge, MA 02142.

This book was set in Adobe Garamond and Rotis Sans Serif by Graphic Composition, Inc. Printed and bound in the United States of America. Printed on recycled paper.

Library of Congress Cataloging-in-Publication Data

Ben-Joseph, Eran.
  The code of the city : standards and the hidden language of place making / Eran Ben-Joseph.
    p.   cm. — (Urban and industrial environments)
  Includes bibliographical references and index.
  ISBN 0-262-02588-4 (alk. paper) — ISBN 0-262-52445-7 (pbk. : alk. paper)
  1. City planning—Standards. 2. City planning—Standards—United States. 3. City planning and redevelopment law—History. 4. City planning and redevelopment law—United States.
I. Title. II. Series.

HT166.B3974   2005
307.1'216—dc22

                                                          2005052228

10 9 8 7 6 5 4 3 2 1

Contents

Preface

In the late 1980s, just out of graduate school and in our newly established consulting firm, my partner Izy Blank and I received a phone call offering the ultimate business opportunity. The developer on the other end of the line was asking for our services to plan and site 650 single-family homes on the outskirts of Tel Aviv. With little business experience but with freshly minted academic enthusiasm and dreams of planning utopias, we were out to change suburbia. No more cookie-cutter homes, rectangular lots, and standardized street patterns. We envisioned a new type of development, one of continuous interplay between space and people, where boulders, ravines, fauna, and flora were interrelated with the human habitat. The reality of development hit us in the face. Minimum lot sizes, setback requirements, right-of-way regulations, roadway-width standards, fire-truck access codes, maximum-density allowances, and so on could have only one outcome: things would be the same as they had always been.

Persistence and youthful disregard for profit making eventually resulted in small victories. Increased densities, as well as shorter, narrower, and more permeable streets flanked by an array of lot sizes, resulted in a more habitable place. Yet the realization that standards and codes are what drive design outcomes lingered. I became interested in discovering why the design process and built environment had come to depend on these criteria and regulations. My journey of discovery started in two other books, *Streets and the Shaping of Towns and Cities* (coauthored with Michael Southworth), and the anthology *Regulating Place* (coedited with Terry Szold). The journey continues with this book. It is by no means the end of the voyage, but another point of departure for those interested in the diverse forces affecting the form and shape of places.

Any product of a long journey owes a great deal to others. In writing this book I have been extremely fortunate in being able to work with friends and colleagues, students and peers of keen intelligence, questioning minds, and profound passion for improving the built environment. Among my colleagues, I am most thankful to Sam Bass Warner Jr. for his long and patient assistance in sorting out my ideas through the logic and directness of his thoughts. In preparing the manuscript, I have found his comments truly welcome, and his encouragement treasured. To Terry Szold I wish to express my appreciation and gratitude for sharing a path of inquiry and reflection while providing me with her indispensable insights. I thank the faculty of the City Design and Development Group at MIT, particularly Dennis Frenchman, Larry Vale, John de Monchaux, Mark Schuster, Anne Spirn, and Julian Beinart, for all of their support, assistance, and countless inspiring conversations. Hiroshi Ishii and the students of the Tangible Media Group of MIT Media Lab, you deserve much recognition for your central role in my work on urban simulation. I am also grateful to Jon Schladweiler, the historian of the Arizona Water & Pollution Control Association, for sharing his vast collection on the history of sewage systems.

I must acknowledge the role of the many MIT students who helped me along the way—in particular, Will Bradshaw, Marissa Cheng, Thayer Donham, Aaron Koffman, Michael Marrella, Tell Metzger, Kath Phelan, Jason Schupbach, Maggie Scott, and Sarah Williams. To two of my students I owe thanks for sharing with me their research work: Jeff Rapson for his study of the communities along the Carolina coast and Michael Sable for his data on the use of technology in urban planning. To Jon Fain, I wish to express particular thanks for editorial assistance and thoughtful suggestions. Many thanks also go to Clay Morgan and others at The MIT Press, and the series editor, Bob Gottlieb, for their receptiveness, support, and helpful comments.

I am also grateful for research grants and other assistance from MIT and the Lincoln Institute of Land Policy.

Portions of this book are related to research published as articles in the *Journal of the American Planning Association, Journal of Planning and Urban Development, Journal of Urban Design, Journal of Planning Education and Research,* and working papers by the Lincoln Institute of Land Policy.

Finally, I wish to express my appreciation to my family, who saw me through this project. Thanks go to Eli, Oren, and Liam for your youthful enthusiasm and mature thoughtfulness, and especially to Holly—without your unconditional dedication, encouragement, and assistance this book would have never come to fruition.

Prologue: Four Tales

South of the mighty Amazon River, deep inside the misty rainforest, a new stretch of unpaved dirt road is about to change this mystical landscape forever. As has happened throughout the region, its completion will open the forest to colonizers attracted by government support and subsidies. Slashing and burning, they will systematically convert primordial jungle into geometric plots crisscrossed with wide streets and lined with pitched-roof homes—places built in the image and model of their suburban counterparts in North America. (*Source:* Courtesy © Bettmann/ CORBIS)

Knee deep, and spattering in the swamp's murky green water, Tuan Nguyen—a future suburbanite—proudly points to the pristine Vietnamese landscape. Here a new vision for Vietnam emerges. It is a vision of progress and advancement, an idea built on images and codes taken straight from the American suburban dream and its engineering "cookbook." Wide, clean cul-de-sacs are replacing the memory of the curving streets of local villages. Generous turning radii make it easy for automobiles to turn corners, rather than accommodating the pedestrian. Ample setbacks and uniform lot widths ensure that intimacy and diversity are replaced by uniformity. All are constructed according to imported codes and standards, transforming a unique setting into an undistinguished place. (*Source:* Courtesy © Annette Kim)

In the rolling hills of the California coast, developer Jeff Slavin knew that he would have to resort to some groundbreaking solutions if he was going to provide affordable detached housing in the booming real estate market. Slavin saw just two alternatives: reduce unit sizes or increase unit density. Choosing the latter, he submitted an innovative single-family project plan, at 9.5 units per acre, for approval to the local planning authorities. It was rejected. The streets were not wide enough, side and front setbacks were not sufficient, and it proposed too many units clustered together. While acknowledging the benefits and innovation of the plan, the local planning authority could not allow such modifications under their current single-family subdivision standards. Slavin was faced with two options, either initiating the lengthy process of pursuing a variance from applicable rules, or opting for a multifamily plan. He chose neither course of action, and instead solved the problem by fencing and gating his project. By creating a private, gated development, Slavin cleared the local government of legal responsibility for public roadways and infrastructure, allowing him to introduce innovative spatial layouts and 144 marketable single-family homes, but isolating his development from many other community neighborhoods. (*Source:* © Eran Ben-Joseph)

Amos Byler, of Chautauqua, New York, is Old Order Amish. In his uncompromising world, very little changes, not the horse and buggy he rides in, nor the hat he wears, nor the home he lives in. Amish standards dictate everything from the plumbing (gravity fed, cold water only) to the oil lamps used in place of electricity to the size of window openings (5 square feet). It is those windows that have suddenly thrust this small Amish community, and town officials, into an uncomfortable clash with the state. The problem is a new state code that requires a minimum opening for bedroom windows of at least 5.7 square feet to ease the escape of residents during a fire and to facilitate access by rescuers. While the sources of the state's dimensions are shrouded in obscurity, there is no ambiguity that such codes disregard local practices and traditions. And since the state is unwilling to compromise over 0.7 square feet, the Amish families may move elsewhere to the chagrin of the townspeople. (*Source:* Courtesy © Joe Liuzzo)

Introduction:

Standards and Rules in Shaping Place

All bad precedents began as justifiable measures.
—Gaius Julius Caesar

No single person dreamed of constructing a sprawling monotonous suburbia in the jungles of the Amazon, or decided to spite Jeff Slavin for his desire to offer denser and more affordable housing. It was the codes and the standards. Across the globe, communities are shaped by standards and codes that virtually dictate all aspects of urban development. Simple standards for subdividing land, grading, laying streets and utilities, and configuring rights-of-way and street widths may seem sensible and insignificant, but because they have been copied and adopted from one place to another, they have an enormous impact on the way our neighborhoods look, feel, and work.

Today's regulations represent the sum of decades of rules designed to promote particular practices. Originating in the desire to improve conditions in urban areas in the late nineteenth and early twentieth centuries, standards became the essential tool for solving the problems of health, safety, and morality. Assuming controls over neighborhood patterns and form, standards shaped the largest segment of urban development in twentieth-century America: the suburbs.

Because so much has been built according to these dictates, the accumulated rules now have the force of universal acceptance—standards have become the definers, delineators, and promoters of places, regardless of variations in landform, natural systems, and human culture. Like genetic code in biology, standards are the functional and physical unit of planning legacy, passed from one generation to the next.

At the present moment, the long historical trend of regulating city building has reached a critical juncture. The expanded application of alternative development regulations and improved development outcomes, such as new urbanism, reflect a kind of societal learning that has resulted from the variety of failures associated with conventional standards. A fresh set of choices is now available, driven by local empowerment, adaptation of place-based guiding principles, and renewed interest in urban form and design. Regulations can continue to accumulate, piling up ever more uniform rules as government and professional inertia carries them onward—or they can evolve, causing a shift in emphasis toward site-specific and localized physical design.

This book reviews the history that brought the modern city and its suburbs to this decision point, and it explores the alternatives now appearing on the urban and suburban landscape in which what is appropriate to be designed and built is found in the facts of cultural distinctiveness and what is normal given the circumstances of place. The intent is not to champion the abolition of regulations or advocate the elimination of all controls or government interventions, but rather to illustrate their evolution and ongoing contemporary effects, and to encourage change where and when needed.

This is a critical point. There is extensive literature by various experts that deals with the numerous forces affecting urban form. From *De architectura* of Roman architect and engineer Marcus Vitruvius Pollio to the contemporary writings of designers such as Christopher Alexander's *A Pattern Language,* Rob Krier's *Urban Space,* and Rem Koolhaas's *Mutations,* many authors analyze and draw normative lessons from cities' shapes and configurations. Historians, geographers, and urbanists such as Camillo Sitte in *Der Städtebau,* Lewis Mumford in *The City in History,* James Vance in *This Scene of Man,* and Spiro Kostof in *The City Shaped* endeavor to understand the form and shape of our cities in their historical and morphological context. Other scholars reflect on the shaping of cities through the eyes of politics and social dynamics. These writings include *Bourgeois Utopias* by Robert Fishman, *Architecture, Power and National Identity* by Lawrence Vale, *La règle et le modèle* by Françoise Choay, and *The Power of Place* by Dolores Hayden. Markets, finance, and wealth inspired nineteenth-century books such as *The Isolated State* by Johann von Thünen, and in the twentieth century, Walter Christaller's *Central Places* and William

Alonso's *Location and Land Use,* which model and predict cities' form on economic and fiscal forces.

In writing this book I have sought to provide neither a survey of these vast and diverse forces shaping cities and towns nor a comprehensive description of their resulting plans. Instead my purpose has been to search out and reveal an important aspect of urbanization: the evolution and role of rules and codes used by societies to create and transform their surrounding physical fabric, and by doing so to provide direction for the design of future residential subdivisions.

Standards and Their Scope and Influence

In the realm of urban planning, standards are extensively used to determine the minimal requirements in which the physical environment must be built and must perform. But they are also seen as the legal and moral instruments by which professionals can guarantee the good of the public. This intent is apparent in the regulation and control over the design and planning of communities and subdivisions.

The past two centuries have been marked by a sustained effort to bring order and safety to the city building process. But what began in the early nineteenth century as a few local and national regulations throughout the United States and Europe is now a worldwide effort toward standardization.

The scope of standards encompasses many different elements of urban and suburban infrastructure. Their influence emanates from and is applied at different levels of government—local governments define land-use controls, building codes are often nationally determined, and state and national environmental legislation affects local development practices. Professional associations and government endorsements have converged to give standards their reach and power. Methodical administration of public works, the centralized supervision of land development, and the rise of the engineering and urban planning professions have established design standards as absolutes.

Certainly, development standards ensure a certain quality of performance, as do many construction standards that are designed to protect our health and safety. However, local governments often automatically adopt and legitimize these standards to shield themselves from lawsuits and from responsibility in decision making. Financial institutions and lenders also are hesitant to support a development proposal outside the mainstream, particularly when it does not conform to established design practices. This happens despite the fact that a mainstream, by-rote solution may be less desirable in its results than a new and creative approach.

Little has been written regarding the actual physical impact of standards on the built form. This may be due in part to the nature of their format, their complex array of criteria, and/or their perplexing idioms. Existing literature in the area addresses various aspects of the subject either as isolated case studies, involving such topics as building codes, or in general terms, focusing for example on the economic impact of standards and regulations on infrastructure development. There is little discussion of the reasons for their widespread adoption in the realm of city planning and design. The lack of attention may also reflect a possible underestimation of the overall influence of standards on form and spatial quality.

Just as genes are obscure, difficult to trace, and often misunderstood as to their impact on bio-organisms, so too are the influence of design standards and their impact on the built environment opaque. The incremental nature of standards reduces each requirement to a singular, discrete mandate. As such, the influence of each standard may seem relatively minor by comparison to the wealth of other variables that are part of the process of urban planning and development. However, when viewed in their totality, their cumulative force has a tremendous effect on the design of places that thus far has been unnoticed.

Today, communities face problems that have arisen because standards intended for health and safety have become disconnected from the original rationale for their existence. These disconnections have overtaken many standards and regulations because the standards have failed to be responsive to their negative impact on the natural and human environments. Residential street standards offer a good example.

Originally intended to afford ease of movement and effortless driving, right-of-way widths and turning radii for these local streets have grown to excess. For the last six decades, such standards demand 50 to 60 feet for the rights-of-way, 36 feet for driving lanes, and a 50–70 foot radius for cul-de-sacs. These standards not only encourage and allow for high-speed driving in residential neighborhoods, but also consume and pave over much desirable open land. In a typical suburban subdivision, with 5,000-square-foot lots and 36-foot streets in a 50-foot right-of-way, street area amounts to approximately 30 percent of the total development. When typical 20-foot driveway setbacks are included, the total amount of space earmarked for cars and driving reaches about 50 percent of the development.

Such land-consumptive requirements often prevent unconventional suburban design practices, such as more densely built developments. As a consequence, opportunities for adaptation and innovation are frustrated and alternative development proposals and building experiments that violate existing standards but might be of great service in creating desirable and sustainable communities are lost.

Revealing the Hidden Language

This book places current planning issues in the historical context of rule and code making. It offers a narrative that takes the reader through the historical evolution of design standards, to an examination of the effect of design rules and codes on the built environment, to a consideration of future directions in the shaping of the regulatory template and place making. Particular attention will be given to residential subdivisions, especially the current template for this type of development that is spreading across the world.

Central to the issue of standards and place making are three sets of questions:

1. *Questions of origins and diffusion*   How were design standards first generated and adopted? Why, when, and how did urban planning and urban design become dependent on codes and standards? And how are they disseminated, practiced, and enforced?

2. *Questions of performance and outcome*   What forms of development have resulted from the codes and standards, and what are their deficiencies? What impact do urban standards have on social and environmental conditions? What are the consequences of the growing uniformity of design standards? How are design standards viewed by those who administer them and by those who must abide by them? And what criteria should be used to measure success, and to determine if standards should be changed or eliminated?

3. *Questions of transformation and opportunities* What are the implications of restructuring design standards? What processes and tools can foster change? Can technological innovations and new forms of information delivery and computing manipulation interfaces create a flexible and more open approach to urban regulations and the application of standards?

To provide answers to these questions, the book is organized into three corresponding parts. Part I, "The Rise of the Rule Book," sketches the historical context and framework in which urban standards and norms have evolved. The chapters in this section describe the factors that have played a major role in the development of standards for urban places. They tell the story of standards and early civilizations, the rise of law and public order, and the establishment of planning and design disciplines and their technical applications. This section begins with chapter 1, "Holding the Commons," which discusses some of the earliest forms of urban standards, those that were characterized by systems of rules based on the power of a sovereign authority, often set apart by divine right and top-down restrictions.

Urban standards have also been shaped by the establishment of professional disciplines and their specific paradigms of practice. Chapter 2, "Experts of the Trade," describes the role of professions that shape urban form, particularly that of land surveying. As part of their early organization, these disciplines had to endorse and apply specific paradigms. This mode of practice enabled the groups to consolidate their positions and define themselves as experts. Thereby the mastery of professional knowledge restricted the role of outsiders to that of uninformed participants with no authority to question professional solutions.

The rapid pace of urbanization in the nineteenth century brought forth an environmental chaos that was linked to the social problems of urban life. At the time congestion, overcrowding, and deteriorating sanitary conditions were believed to cause social and moral degeneration. The desire for better living conditions prompted interventions by public authorities. These interventions provided the foundation for the form and shape of new neighborhoods to come. Chapter 3, "Neighborhoods Developed Scientifically," describes some of these early regulations.

Part II, "Locked In Place," illustrates how standards are forcing an exclusive planning process and limiting alternatives for physical design. The chapters in this section provide examples to demonstrate how standards have contributed to the shaping of neighborhoods and cities. They also cover attitudes among both the private sector and public agencies about the extent, nature, and effect of standards on development in the United States, as well as the influence of these standards on development overseas.

Chapter 4, "Sanitized Cities," illustrates how past technological choices have shaped current planning practices while often discouraging change. Like many other aspects of city infrastructures, sewer-system standards, for example, are so entrenched and widely accepted that alternative planning, sizing, and locating of the systems are seldom considered. Should standards be based on technologies within the current paradigm, or should they be based on the long-term goals decision makers want to achieve? Next, chapter 5, "Regulating Developers," evaluates the impact of standards and regulations on the design of residential developments. It assesses common attitudes and perceptions, and identifies the issues that members of the housing industry and its regulatory agencies feel are affecting housing development.

Concluding this part of the book, chapter 6, "Second Nature," discusses the impact of standards on the landscape and its natural systems. One of the most troublesome stages in the site-development process involves the clearing of existing vegetative cover. The desire to cut costs by executing massive grading with heavy equipment often results in complete alteration of the landscape and degraded environmental conditions. Local governments have generally recognized the consequences of such practices and many have adopted standards for this development phase. Yet these regulations are not only poorly implemented and enforced; they are seldom revisited or revised.

By becoming more aware of how standards have evolved over time, and how profoundly they affect our places of living, many seek a more equitable planning process, and look to improve the design of our built environment. Part III, "Altering Inherited Traits," takes as its focus the likelihood of a paradigm shift. It points to new conditions that ultimately must lead to change in the application of rules and standards to development. These new forces can be seen in the growing environmental awareness of the public and private sectors, as well as in new partnerships between the two in the development of sustainable technologies, and in the introduction of new tools for facilitating public participation in place making.

Chapter 7, "Private Places and Design Innovation," describes how private developments are pushing the sustainability envelope, protecting their environmental resources in an effort to increase marketability and financial return. This has resulted in a transition from traditional individual ownership of property to collective governance. This fundamental change not only represents a shift in neighborhood governance, but most significantly calls forth a change in the physical character of residential development—often in the form of innovative spatial and architectural layouts, in some cases as a result of unusually sensitive environmental design. It also creates a de facto deregulation of municipal subdivision standards and zoning. Many of the ecological concepts of these private communities can be applied to the broader housing market, given the consumer's willingness to pay for environmental quality, or by offering public incentives to fill the economic gap.

While two-dimensional maps, charts, and diagrams to computer models allow "experts" to explain their designs and planned interventions more clearly than ever before, few platforms exist that allow immediate, real-time, and seamless changes in response to public or professional input. New and promising technologies are discussed in chapter 8, "Technogenesis and the Onset of Civic Design." These innovations have the potential to create a paradigm shift in the application of design standards to the process of place making. These new systems could be used not only as tools for design professionals but also as an interactive application to enrich communication and learning within the design process. The integration of such envisioning tools will allow for better professional judgments while incorporating various stakeholders' expectations.

The concluding chapter—chapter 9, "Places First"—calls for a design methodology consistent with and based on site-specific context. Only local conditions and physical context should provide the threshold for the formulation of standards and codes. Regulations should be place-based, emphasize details, and be buttressed by public approval. As more communities wrestle with problems due to uncontrolled growth, environmental pollution, and failures of the existing infrastructure, they are likely to take a stronger interest in their local power. Thus the possibility for communities to establish their own initiatives for localized place-based standards can be realized.

No doubt, regulations will continue to exert influence and shape the built form of the global landscape. The future of the regulatory shaping will inevitably evolve from the templates we have used in the past. But if regulations are too inflexible to allow for innovation, then perhaps we must work to see that they are changed.

Above all, planners and designers must take formal stands against the adoption of rules that perpetuate mediocre development outcomes. There should be a willingness to test standards, not only in relation to preventing harm or preserving property value, but in relation to their impact on the physical form of communities.

Standards are the source of how communities are designed and built. They define how places *can* and *can't* be developed, and how controls shape the physical space where we live and work. It is the aim of this book to help unmask and explain this relationship, for though standards will continue to exert their influence on the shaping of our towns and cities, we must not allow them to prevent excellence and innovation in our quest for better places.

Part I

The Rise of the Rule Book

1

Holding the Commons

> If a builder builds a house for someone, even though he has not yet com-
> pleted it; if then the walls seem toppling, the builder must make the walls
> solid from his own means. If a builder builds a house for someone, and
> does not construct it properly, and the house which he built falls and kill
> its owner, then the builder shall be put to death.
>
> —Hammurabi, King of Babylonia, 1780 BC

Black Rock, Nevada

For seven days each year, the dry, desolate lake bed of Nevada's Black Rock
desert is filled with life. In what has become an annual pilgrimage, tens of thousands of people flock into this remote part of the continent to build their temporary camps and sculpt artifacts that are to be ultimately destroyed in bursts
of fire. At the center of this ephemeral city stands its focal effigy—the wooden
statue of the Burning Man.

For a gathering with no rules and regulations, communal etiquette and
social norms dictate conduct and the making of place. There is no commerce,
and no driving, except to park at one's campsite along the circular streets. Although never planned, the circle shape of this temporary city grew from an in-
stinctive urge to round the gathering in the boundless space of the desert. The
circle of camps gradually developed into a civic hub that centralized city services, provided a social gathering place, and created a prominent landmark with
its central effigy. Four avenues were added to the circle to indicate the cardinal
directions. A second circle, called Ring Road, surrounded this civic center, and
a no-camping zone called No Man's Land was established to preserve the view
of Burning Man.

As the "city" grew, health, safety, logistics, and environmental degradation required centralized intervention, rules, and regulations. By 1998, seven years after the first city had been constructed in the desert, planning rules and physical plans were enacted. The city arranged around itself a geographic center formed by the location of the Burning Man. From this spot, an arc that defined the curve of Black Rock City's concentric streets was surveyed and plotted. These arced streets were subdivided into blocks by a series of radial streets, like spokes projecting outward from the hub of an enormous wheel.

Zoning soon followed. Special areas were designated for walk-in camping, law enforcement, sanitation, public works, and even an airport. Growth and expansion became an issue. Strict boundaries had to be drawn, yet flexibility for growth had to be accommodated. "As our plan has grown," the organizers acknowledged, "we have learned how to differentiate and separate various specialized and potentially conflicting uses. This very much involves an empirical study of our social needs as they've naturally emerged from an increasingly sophisticated social reality."[1] (See figure 1.1.)

The Norms of Early Human Settlement

The ephemeral city of Black Rock reenacts the story of many cities through the ages. It is a small example of how a degree of social order evolves from the ground up, how social norms and conduct are often the cornerstone by which communities are built and constructed, only later to be transformed into rules and regulations often imposed by a central authority. Perhaps the special quality of its current rules resides in its circular layout, an uncommon form for most towns and cities. Thomas Hobbes, who stands at the fountainhead of modern political thought, often argued that the human "state of nature," or natural condition, was one of chaos, disorder, and conflict. Hobbes suggested that to avoid this naturally occurring anarchy, the mighty leviathan of the state must step in to impose orders and rules. "For the laws of nature," Hobbes wrote, "[such as] justice, equity, modesty, mercy, and in sum, doing to others as we would be done to, of themselves, without the terror of some power to cause them to be observed, are contrary to our natural passions, that carry us to partiality, pride, revenge, and the like. And covenants, without the sword, are but words and of no strength to secure man at all."[2]

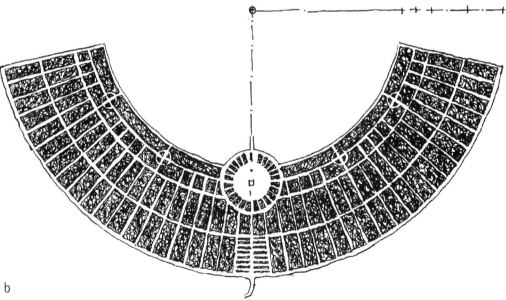

b

Figure 1.1
Black Rock City, Nevada, started as an informal gathering constructed according to flexible norms and social conduct. Through the years rules replaced norms to shape the city according to exacting design standards. (*Sources:* Photo, courtesy © Brad Templeton (a); Plan; Eran Ben-Joseph (b))

Indeed, centralized rules for the built environment are as old as the rise of the city-states in the valleys of the Indus, Nile, Tigris, and Euphrates. As these cities expanded to become complex systems, laws became codified and social norms were turned into standardized practice. Regulating production as well as controlling and administering land, always a precious commodity, gave rise to governing techniques. The Egyptians, subjected to the annual flooding of the Nile, created a benchmark system to readily reestablish property lines once the water receded. Clay tablets and artistic representations dating back to the Sumerian culture of 1,400 BC show records of land measurements and plans of agricultural and built areas. In China, a unified land-measurement system was enacted by the self-styled first emperor Shin Huang-Ti in 221 BC. Yet it seems clear that these technical codes bore a complex relationship to the communities they attempted to order: they mixed government authority with a recognition of the customs of the place.

While rules and norms of social conduct are commonly mentioned in ancient treatises, little is actually known about rules for city planning. Evidence from archeological excavations such as those in Kahun and Tel-El-Amara in Egypt suggests that these towns, built 3,000 to 4,000 years ago, were laid out in a formal pattern with straight streets and small blocks filled with dwellings abutting each other. Wider avenues connected important civic buildings, and a clear differentiation in the density and size of the housing is associated with each sector of the city. Other excavations such as those in the Indus Valley, at Mohenjo Daro and at Harappa in Punjab, show that the valley had cities planned and built in rectangular blocks lined with two-story houses along broad, straight streets. Many of these streets and buildings were drained using covered sewers that, at regular intervals, led into buried cesspools. Such evidence of the pervasive use of orthogonal arrangements tells of an enduring continuity between farm, village, and city. The farmer's custom of long narrow fields and right-angled boundaries carried over easily into streets and squares. The innovative geometry of towns and cities lay more in the problems of water supply and drainage than in right-angled street corners.

Indus Valley

The Indus Valley civilization also offers the earliest known written documents on city law. In the religious writings of the Vedas, which some scholars believe may date back 4,000 years, strict rules and codes about planning cities, laying streets, and constructing buildings were inscribed.

The following typifies the breadth and extent of the Indus Valley rules and standards:

On site suitability for building construction:
Dig out a pit one cubit deep in the ground and again return the earth into it. If the earth more than fills up the pit, then the ground is good; if it is just sufficient then it is middling or indifferent; while if it falls short, the ground is bad. The good and indifferent varieties are acceptable, but the bad should on no account be accepted.[3]

On laying out a town and its streets:
First lay out the town and then only plan the houses. Violation of this rule portends and brings evil.

First plant the trees and erect the premises thereafter: otherwise they will not look graceful and seemly.[4]

On houses and their relationships:
The houses of Brāhmans should be chatuhśālā; that is, they must occupy the four sides of a quadrangle which is an open space in the center.

The imperial palaces should be raised to eleven stories; the building of the Brāhmans to nine stories; those of ordinary kings to seven stories.

As far as possible, the height of buildings in the same street should correspond, that is to say, one should not be lower and another higher.

A deviation from the fixed measurements of lengths, breadths, and height of the respective buildings of the different classes of peoples is not conducive to good and should not be made.

Outside the house and touching it there should be planted a foot-path (vīthikā). All houses should face the royal roads and at their back there should be vīthīs or narrow lanes to allow passage for removal of offals and night-soil.

Not encroaching upon what belongs to others, new houses may be constructed.

From each house a water-course of sufficient slope three pādas or one and a half aratnis long shall be constructed that water shall either flow from it in a continuous line or fall from it into the drain. Violation of this rule shall be punished with a fine of fifty-four panas.

Between any two houses, or between tile extended portions of any two houses, the intervening space shall be four pādas or three pādas (feet).

The owners of houses may construct their houses in any other way they collectively like; but they shall avoid what is injurious. Violation of this rule shall be punished with the first amercement.[5]

Many of these design rules protected property from harmful actions by abutters, especially those that required building setbacks to preserve light and air, and to allow adequate drainage. Minimum physical-development standards such as construction of sidewalks and alleyways, and the allocation of particular building types to specific areas of the city, resemble contemporary American zoning practices. However, these rules depended as much on the goal of maintaining the social hierarchy as they did on the imperatives for safe modes of settlement.

Social stratification led to the segregation of classes and professions into specific wards in the city. In his book *Town Planning in Ancient India*, Binode Behari Dutt explains: "When the principal streets called Rājapathas are laid out, the whole city area is divided by them into 'grāmas' or muhallas, i.e., wards. . . . Distribution of professions and casts as well as allotment of sites were made entirely with reference to pada-vinyāsa, a pada or block being set apart for a caste or profession."[6]

China and Japan

Urban planning standards that emphasized rectangular subdivision to maintain social rank and function also played an important role in the ancient cities of the Far East, particularly those of China and Japan. One of the oldest descriptions of regulation for the construction of cities and their architecture is found in the *Zhouli* (The Rites of Zhou), one of the thirteen Confucian classics dating from the Chinese Zhou Dynasty of the second century BC. A famous passage describes the laying out of King Cheng's imperial city of Luoyi:

> The jiangren [architect, builder] builds the state, leveling the ground with the water by using a plumb-line. He lays out posts, taking the plumb-line (to ensure the posts' verticality), and using their shadows as the determi-

nators of a mid-point. He examines the shadows of the rising and setting sun and makes a circle which includes the mid-points of the two shadows.

The jiangren constructs the state capitals. He makes a square nine li on each side; each side has three gates. Within the capital are nine north-south and nine east-west streets. The north-south streets are nine carriage tracks in width. On the left (as one faces south, or, to the east) is the Ancestral Temple, and to the right (west) are the Altars of Soil and Grain. In the front is the Hall of Audience and behind the markets.[7]

These passages illustrate the fundamental principles and components of every Chinese imperial city. Although no specific dimensions were required (except for the width of the north-south streets), strict divisions along geometric right-angle lines and specific usage were designated. The strong emphasis on geometric gridlike patterns was derived from religious and philosophical influences, as are found in Confucianism. For example, Confucianism's emphasis on the imperial system of power and on the centrality of the emperor is reflected in the ways the palace is placed in the city's center. The square-shaped, symmetrical city, with houses located in different wards, often according to social ordering, was designed to reinforce the vision of the imperial core as the appropriate moral focus for society and daily life.

Other influences on Chinese city form are attributed to hexagrams (geometric figures composed by six lines) associated with symbols of fortune. Described in the *I Ching* (Book of Changes), the sixty-four hexagrams are formed by joining in pairs, one above the other, basic broken and solid lines each with its own meaning. The influence of the *I Ching* has often been associated with the laying out of major streets, channels, and the division of wards within the Chinese city.

The capital city of Chang'an is said to have been constructed with strict geometric lines based on various hexagrams. Twenty-five streets crossed each other, forming a checkerboard pattern. Along the horizontal (east-west) streets, special demarcations corresponding to the *I-Ching* hexagram were constructed. The palace was built on the second line derived from this hexagram because "the dragon is perceived as an open space; it is advantageous to visit a great man."[8]

Chang'an had a great influence on subsequent urban planning. Other East Asian cites often modeled their capitals after this particular prototype, adopting the rectangular grid patterns for streets and city blocks and replicating the location of the palace and the various wards. Examples include Kyongju of Shilla in Korea, and Nara and Kyoto in Japan. Nara, the capital of Japan from 710 to 784 AD, adopted the Chinese grid-pattern system (figure 1.2).

Although no specific records of regulations exist for the building of Nara, an archeological discovery in 1979 of a local tomb indicated that the city was carefully planned and administered. An inscription found in the tomb specified that it was a burial of "Ô-no-Ason-Yasumaro, a resident (in the city-block) of the fourth Jô (row) and fourth Bô (column) in the Left City (of the Capital), at the junior fourth-rank lower in the Court and the fifth Order of Merit, died on the 6th day, 7th month of the year of Gui-Hai (the 60th year of the Chinese zodiac cycle) (August 11, 723)."[9] Thus we can tell that every place in the city was specified according to its wards and cross-street numbers.

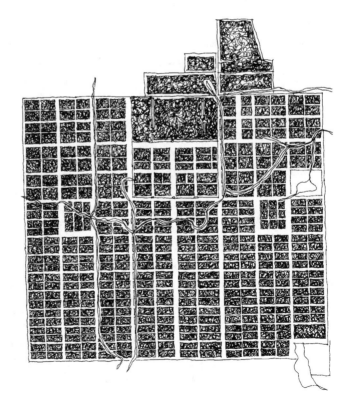

Figure 1.2
Although no specific dimensions were required in the planning of Chinese Imperial cities, as can be seen in this plan of Chang'an (modern Xi'an), strict divisions along geometric right-angle lines and specific usage were designated. The strong emphasis on geometrical gridlike patterns was derived from religious and philosophical influences such as those found in Confucianism. (*Source:* Eran Ben-Joseph)

Greece

The need to control the growth of cities and to maintain public order was also part of the growing culture along the Aegean Sea. Aristotle and Plato tell us that around this time of expansion (350 BC), Greek cities began to pass bylaws relating to the policing and the securing of the public order of the markets, the agora (public square), and the streets. These bylaws were enforced by special officials—the Agoranomi to control the markets and the agora, and the Astynomi to control the streets.[10] Plato also provides us with the specific laws governing the construction of new cities. In his *Laws* he writes:

> It would seem that our city, being new and houseless hitherto, must provide for practically the whole of its house-building, arranging all the details of its architecture including temples and walls. . . . When we come to the actual construction of the State, we shall, God willing, make the houses precede marriage, and crown all our architectural work with our marriage-laws. For the present we shall confine ourselves to a brief outline of our building regulations.
>
> The temples we must erect all round the market-place and in a circle round the whole city, on the highest spots, for the sake of ease in fencing them and of cleanliness: beside the temples we will build the houses of the officials and the law-courts, in which, as being the most holy places, they will give and receive judgments,—partly because therein they deal with holy matters, and partly because they are the seats of holy gods; and in these will fittingly be held trials for murder and for all crimes worthy of death.
>
> [As to walls,] if men really must have a wall then the building of the private houses must be arranged from the start in such a way that the whole city may form a single wall; all the houses must have good walls, built regularly and in a similar style facing the roads, so that the whole city will have the form of a single house, which will render its appearance not unpleasing, besides being far and away the best plan for ensuring safety and ease for defense. To see that the original buildings remain will fittingly be the special charge of the inmates; and the city-stewards should supervise them, and compel by fines those who are negligent, and also

watch over the cleanliness of everything in the city, prevent any private person from encroaching on State property either by buildings or diggings. Officers must also keep a watch over the proper flowing of the rainwater, and over all other matters, whether within or without the city, that is right for them to manage. All such details—and all else that the lawgiver is unable to deal with and omits—the Law-wardens shall regulate by supplementary decrees, taking account of the practical requirements.[11]

Plato's laws reflected a common understanding of his time and were often applied as actual standards in the planning of cities. Pergamon, in northwest Asia Minor, was one of those cities. Constructed around 300 BC, it was a typical hill town laid out in a semicircle crowning the peak and on the inner slopes of a crescent-shaped ridge. Near the highest end of the ridge, ringing the hilltop, stood the palace, the temples, the agora, and other civic buildings. Lower, beneath the agora, the town covered the slopes, cascading down the mountain. Such an amphitheater form of a hillside town gave the new Greek settlements a more circular plan that was common elsewhere. (See figure 1.3.)

An inscription from Pergamon citing part of the Royal Law provides us with a glimpse of the application of these rules. It calls on the Astynomi to su-

Figure 1.3
Pergamon in northwest Asia Minor (modern Turkey); model of the upper town. (*Source:* Bildarchiv Preussischer Kulturbesitz/Art Resource, New York)

pervise the upkeep of houses. If houses collapsed or were about to fall into disrepair, the owners had to be warned of the consequences. When the owners declined to take action, the magistrates would assume the responsibility for the repairs and then charge for the cost of the work in addition to imposing fines. There were provisions for the maintenance of the city walls and public streets as well as for the elimination of runoff water from one property to another.[12]

It is not surprising that in later years many of these common rules found their way into the bylaws of Roman municipalities. Both the ancient Greek and Roman cultures tried to manifest their high regard for civic life and urban culture by bestowing duties as well as privileges on their citizens.

Rome

Historical evidence of purposeful city planning and design is most plentiful in the ancient Roman period. Of particular importance is the extensive excavation of the intact city of Pompeii. Information also has been gained from traces of ancient city plans in existing Italian cities and Roman colonies such as Carthage in North Africa, as well as Silchester in England.[13]

Roman city planning often followed a systematic layout of a gridlike pattern. Scholars have attributed this pattern to early agricultural land-demarcation practices around Rome. These farming subdivisions, easily measured and controlled by their owners, influenced the laying out of Roman military camps, and eventually shaped the regular forms of Roman colonial towns.[14]

The grid, in turn, was often reinforced by two main streets crossing each other in the center at right angles. Some have argued that these streets, the *cardo* and the *decumanus,* were such an essential and widely used element of urban planning that they were the cornerstones of any new colonial expansion. Yet little written evidence exists to indicate that Romans had to follow strict rules or standards in planning their cities.[15]

Roman writings on surveying, for example, reveal more concern with site conditions such as soils, wind, and light than with the arrangement of the town. These concerns for light, air, and the well-being of citizens generated the few rules that have been documented. In Rome, for instance, various emperors attempted to keep industries out of central areas, to prevent projections of buildings into streets, and to set limits on the height of buildings.

One of the few treatises on the principle of Roman city building is that of architect Marcus Vitruvius Pollio. A first-century BC Roman architect, he was the creator of the original handbook *De architectura libri decem* (Ten Books on Architecture), published around 40 BC. Although the book's primary concern is theories of good architecture, it also offers rules and design standards, as well as glimpses into Roman building practices.

Vitruvius is also concerned with the layout and design of public spaces and streets. He suggests the principle that building height should relate to the width of streets and courts in order to allow for the adequate lighting of interior rooms and the open rear courtyards common to Roman houses. To achieve the ample lighting he mentions, Roman public laws forbade exterior walls of buildings abutting public ways to be more than one and a half feet thick. According to Vitruvius, this regulation attempted to forestall the addition of new stories to existing buildings by means of the widening of the lower walls. When the thickness of supporting walls grew with the increased height of a building, existing streets and rights-of-way became too narrow.

When the tendency to build higher houses left dark narrow passages, insufficient for wheeled traffic, Augustus Caesar (63 BC–14 AD) responded by limiting the height of buildings to 66 feet, or no more than six stories. He also required that new construction along the *decumanus* (processional road) be limited to 40 feet in height, the *cardo* (main road) to 20 feet, and the *vicinae* (side road) to 15 feet. After Augustus, Nero (37–68 AD) stipulated new guidelines for buildings, advocating that they not exceed in height two times the street's width.[16]

Still, it is unclear if these street-related standards possessed the force of law. The few actual rules relating to city buildings, as stated in Roman laws and charters, typically dealt with the unlawful destruction of buildings, and not with the design of streets. For example, the charter of the municipality of Tarentum, during the time of Cicero (106–43 BC), stated that "in no case shall a person destroy, unroof, or dismantle a building unless he is ready to replace it with another building of the same sort, or if he received a special permission from the town council."[17] Such rules were more typical of the time. They were rules and standards aimed at protecting citizens from the damaging acts of others, and, perhaps, at protecting the common value of town properties. Less usual were rules designated to standardize the design of the cities.

Byzantium

Julian of Ascalon, an architect and a native of the Byzantine Palestinian coastal city of Ascalon and a contemporary of the Byzantine emperor Justinian I (483–565 AD), wrote a treatise that is a compilation of construction and design rules that deal with land use, views, house construction, drainage, and planting issues.[18]

The controls on drainage provide an interesting example of an attempt to limit flooding and damage to nearby houses in the typical compact towns of the sixth century. The treatise classified two types of drainage: rainwater and wastewater. Specific dimensions and rules are provided for the laying out of drainage pipes and their connection to a central collection system. These rules were assigned according to the specific material to be used. Any upgrading, such as the use of metal instead of stone, would reduce the required setbacks and narrow the pipe's dimensions because of a greater efficiency, thereby offering an incentive for their use.

Another unique rule in Julian's treatise, rare even in our own times, was the control of views. Situated along the shores of the Mediterranean, and facing a bustling harbor, views from buildings in Ascalon were meant to be preserved. The treatise indicated that new construction could not obstruct a direct view of the water. To provide some guidelines, views were placed in different categories such as foreground, to include the shores, harbor, and anchoring ships; middle ground; and far distance. Owners had the right to complain about construction that blocked their direct view of the foreground features, but not those that blocked the faraway vista. Although no specific building-height requirements were given, setbacks were used as the mechanism to preserve views. (See figure 1.4.)

Julian's treatise, like many from that time, was primarily concerned with maintaining a level of equitability and addressing change. It did so by combining a large measure of performance outcome and a small dose of prescribed rules. This distinction between performance requirements and prescription had an important implication for place making. Performance-based rules tended to allow for freedom with respect to actions and solutions within a framework of established norms. Prescriptive rules were often designed by a central entity in a top-down fashion in a manner that often had little grounding in the essence of place.

Figure 1.4
A unique law in Julian's treatise was the control of views. Situated along the shores of the Mediterranean and facing a bustling harbor, views from buildings in Ascalon were meant to be preserved. Shown is a partial plan of the city as depicted in the Madaba Mosaic. (*Source:* Courtesy Madaba Archeological Museum)

Due to their flexible nature, performance-based codes tended to evolve over long periods of time and were often associated with customary law and the ethical systems and values of a community. Their emphasis on doing no harm to neighbors, rights of original usage, privacy, and obligations concerning environmental degradation, have affected many Mediterranean cities that developed in the Byzantine period.

The range of performance-based codes is particularly interesting as seen in relation to the organic form of the medieval European and Islamic city.

Open-Ended Performance Standards in Islamic Cities

The Sharia—Islamic law based on the Koran—has often been applied in the context of the Islamic city to the regulation of land use, the control of building heights, and the location of doors and windows.[19] Sharia's influence extends beyond law. It encompasses religious, political, social, domestic, and private life. Its social and domestic core encompasses the injunction to avoid causing harm to others and the regulation of public interest (Maslahah).[20]

These regulations and norms addressing the public interest and preventing harm to others can be seen in the control of streets within the city, where the public interest takes precedence over private concerns. Typical of such areas is where vendors are present, or in the *suq* (market). As stated by a jurist, Ibn al-Ukhuwwah, in the twelfth century:

In narrow streets, traders must not set out seats or benches beyond the line of pillars supporting the roof of the suq so as to obstruct the way for passersby. The prolongation of wooden beams and projections, the planting of trees, and the building of benches are forbidden in the narrow streets, the way through the suq being common property through which the public has the right to pass. . . . So also the tethering of animals is forbidden except as required for alighting and mounting. Sweeping refuse into the passageway, scattering melon skins and sprinkling water, which may cause slipperiness . . . these are all forbidden. . . . Water spouts may not be allowed to project from walls so as to cause defilement of the clothes of passersby and obstruct the streets. Rain water and mud must be swept away from the streets and it is the duty of the muhtasibs to compel people to take care of such matters.[21]

Islamic law and rules of the Islamic city show a great deal of adaptation and emphasis on social behavior rather than prescriptive physical regulation. It seems likely that the enforcement of such rules depended more on the customs of the town than on the role of officialdom. Rarely, if at all, does one find physical proportions or specific dimensions for laying streets or constructing housing. Concerns for privacy, for example, did not result in uniform architecture and building layouts. In many cities, such as Medina, various housing typologies were developed, all with their own distinct height and shape. Thus the Islamic city is a unique example of the achievement of urban-form conventions through principles of use rather than specific architectural regulations. (See figure 1.5.)

Maybe the most interesting example of these practices are the rules associated with the fina—the space around entrances and along buildings facing a street. In most Islamic cities, such spaces are utilized by families either for social interaction or for commerce. It is common to see merchants placing their goods on the streets next to their shops or to have families gather to sit and socialize near the entrances of their dwellings. While no strict standards or dimensions define these spaces and although they definitely fall within the public domain, there is a clear acceptance that these front areas are subject to the use of the immediate residents. Jurists writing at the time show a lack of willingness to intervene in regulating these semipublic spaces as long as the owners and abutters are in agreement concerning their use. Malik of Medina, for example,

Figure 1.5
Islamic cities such as
Damascus, Syria (a), and
Kalaa Sghira, Tunisia (b),
are examples of the
achievement of urban-
form conventions
through principles of use
rather than specific ar-
chitectural dimensions.
(*Sources:* Plan, Eran Ben-
Joseph; image, courtesy
© Roger Wood/CORBIS)

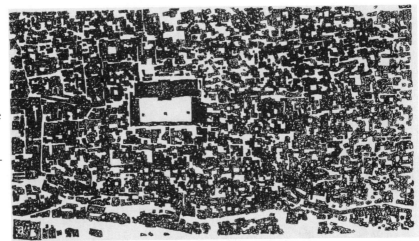

writes that "for spaces of small width, where the least thing posed would hinder the circulation, I think that no one has the right to reserve their use for himself, and that the authorities must intervene; but for those where the width is such that the circulation would not be hindered at all if the neighboring owners utilized them for their own need, I see no harm if the authorities do not intervene."[22] (See figure 1.6.)

No specific dimensions define the fina. Rather it is an imagined space marked by activity and conduct rather than by an exact line. The rules are based more on social behavior and cultural norms than on prescriptive physical regulations. Such social norms suggest that in earlier times, prescriptive, quantitative, and centralized systems of city planning rules were not the sole force behind urban regulations. More likely, systems of controls were based on community norms and conduct, and may have been pluralistic in nature. In the Islamic cities, this type of system shows a great flexibility and adaptability to local physical and social conditions. This state of affairs tended to safeguard and perpetuate the distinctiveness of local buildings and urban design. In *Arabic-Islamic Cities: Building and Planning Principles,* Besim Hakim suggests that this type of control "accords legitimization and protection to a locality's customs and practices and thus contributes substantially to the identity of a place through the individuality of its place-making process and its resulting built form."[23]

Few performance norms have survived the impact of Western styles and fashions, and of Western conceptions of city planning and architecture. In Islamic society, as well as in other traditional cultures, changes have been forcibly and rapidly brought about by colonial powers, and local rulers wishing to modernize have had to do so according to foreign models. The primary victims of those changes have been the traditional norms for the built environment.

Medieval Europe

The design and regulatory practices of medieval European towns may be conveniently summarized in terms of customary practices and an alternative new-town process. After the fall of the Western Roman Empire and the disastrous depopulation caused by the eighth-century plagues, Europe was left with a number of towns planned and walled by the Romans, such well-known places as London and Paris, and a variety of crossroad villages and manors.

Figure 1.6
There are no specific dimensions that define the *fina*. Rather it is an imagined space marked by activity and conduct rather than by an exact line. The rules are based more on social behavior and cultural norms than on prescriptive physical regulations. (*Sources:* MIT Rotch Visual Collection—the Aga Khan Fund)

A typical town of the tenth century might consist of a stockade and some towers constructed to provide a sheltered market and church, and to give some protection to the farms nearby. The essential element lay in the "high street," a more or less straight street than ran from the gate to the market. Here peasants and shepherds as well as a few local merchants and craftspeople met to trade. In such a village the houses stood tightly packed together and the streets ran in a patchwork of branches, each cluster taking its orientation from the shape of the land or the location of a river or stream. There were right-angled intersections, but their presence depended on the local landform.[24]

Population growth and an increase in trade, especially from the twelfth century on, proved the key design and living problem for such towns. The wall stood as a massive public burden, and for the average town dweller, "as long as *his* house and *his* shop, and *his* parish, and *his* church were contained within the walls, then the wall was big enough."[25] This enduring pressure exerted by the town wall called forth the successive alleys and building encroachments that gave the medieval town its notorious reputation for dark, crooked lanes, tortuous passageways, and impassible narrows. The town fathers were forever passing ordinances against encroachments, but to little effect.[26] In 1262, for ex-

Figure 1.7
To accommodate the military necessity of a defensive zone near the city's perimeter, Siena passed a statute in 1262 that required houses along the wall to be fitted with merlons. The illustration shows the city depicted in the Frescoes of the Good and Bad Government Siena, Palazzo Pubblico, by Ambrogio Lorenzetti, 1338. (*Source:* Scala/Art Resource, NY)

ample, Siena, trying to keep a defensive zone near the wall wide and clear for military purposes but also wanting to utilize the space for dwellings, compromised by passing a statute that required houses along the wall to be fitted with merlons.[27] (See figure 1.7.)

A planning process coexisted with the customary incremental growth of medieval towns. The founding of new towns was a part of the resettlement of Europe and the reclaiming of the forest. Kings and great feudal lords established simple rules for the layout of their new towns. A site suitable to be enclosed by a protective wall was selected. Within it, a few carriage-width streets connected the gates to a central market, town hall, and church. A rectangular system of minor streets provided new settlers with frontage lots on which to build their houses. Such planned towns, *bastides,* dotted southern France in the twelfth and thirteenth centuries, and they also were pressed into service on the frontiers of conquest, as they were in Wales and Ireland. Occasionally, too, the older cities imitated these rectangular practices when they extended their walls to add a new quarter, as they did in the new town of Edinburgh and in the lower town of Boulogne-sur-Mer.[28] Indeed, such medieval practices for the layout of towns migrated to the New World in the Spanish laws of the Indies, and in colonial North American experiments with town planning.[29]

Shifting Paradigms

The limited ambition of medieval European regulations—a wall for defense, a few streets wide enough for a wagon, a town square with its market days, and some right-angled streets—left much to the particular site and the customs of the inhabitants. A thoroughgoing command of town layout and building lay far beyond the reach of the medieval tradition. The reconstruction of Riyadh in Saudi Arabia tells a very different story. It reports on the consequences of the abandonment of traditional Arab-Islamic performance-guided norms in favor of imported standardized urban concepts.

In the early twentieth century, a decision was made to relocate all government agencies from Mecca to Riyadh. As part of this plan, the Saudi government decided to transform the small, sleepy residential town into a modern capital built according to Western principles. Setting a network of grided streets and large lots to accommodate new building typologies and construction materials, a clash between tradition and modernity began. (See figure 1.8.)

In contrast to the traditional pattern of narrow and winding streets, the new neighborhoods were orthogonally planned, crisscrossed by broad boulevards and wide streets. Local resources of clay, sun-dried mud bricks, and wooden roofs were replaced by cement, reinforced concrete, and tile. Soon enough these

Figure 1.8
The new plan for Riyadh standardized the rectangular lot with its strict setbacks, side-easement configuration, and the modern villa. The ease by which new land could be subdivided, developed, and allocated was in keeping with the needs of a fast-growing economy. It provided policymakers with an image of urban progress and modernization built on Western standards and juxtaposed against the perceived chaos of the traditional city. (*Source:* Courtesy Saleh Al-Hathloul, MIT Rotch Visual Collection—the Aga Khan Fund)

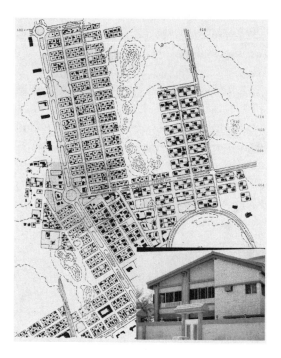

practices were set in stone through the establishment of codes and standards as well as government enforcement. Sought-after foreign experts and consultants further entrenched this new vision. Doxiadis Associates, for example, an urban planning and architecture firm based in Athens, Greece, was hired to develop the new grand vision for the city.

The plan, devised by the foreign experts, and ultimately approved by the Saudi government in 1973, stated some important ideals, yet used the wrong tools for their implementation. In part that plan hoped to capture the essence of the existing vernacular city. It stated that "in any new comprehensive planning legislation, special building rules and regulations should be drafted to ensure the maintenance of the basic principles of local architecture (i.e. internal courtyards, etc.) without necessarily mimicking old and obsolete architectural forms and construction techniques."[30] While novel in its efforts to preserve the traditional spirit, the plan's actual implementation devices ensured a different outcome.

By means of specific engineering design standards and regulations, the scale and character of the city were transformed to resemble a contemporary Western suburb. New design standards—for example, specifying minimum lot size, the bulk and height of buildings, and street configurations—guaranteed that any construction along traditional norms would be unattainable. Minimum lot sizes and setback requirements forced the utilization of square lots for land subdivision. Lots were not defined by square area measurements but by specific length and width dimensions. Such specifications were exacerbated by government distribution of plots through various housing programs and their call for the various jurisdictions to follow the prescribed rules. The following excerpt from an official Saudi government directive typifies such enforcement: "The government of his Majesty the King is deeply committed, on every level, to finding suitable solutions in order to minimize the housing crisis. . . . Thus we urge all municipalities to undertake a search for government land suitable for building within the limits of the master plan of their cities and, in cooperation with the town planning office, to subdivide these lands into residential plots with an area in the limits of 20×20 meters each. And then to distribute them to fellow citizens (especially those of limited income), either as grants or by selling them according to the rules designed for this purpose."[31]

The standardization on the rectangular lot, with its strict setbacks and side-easement configuration, has also entrenched the street grid as a desired pattern. The ease by which new land could be subdivided, developed, and allocated appealed to the needs of a fast-growing economy. It also provided policymakers with an image of urban progress and modernization built on Western standards and juxtaposed against the perceived chaos of the traditional city.

The dictates of rectangular lots, prescriptive setback requirements, and low-density rules have also brought about a change in housing typologies. Traditional tightly packed and stacked dwellings were replaced by the single freestanding house. Packaged as part of the new modern city image, the single home became the prominent dwelling alternative, by law if not by choice.

Order and the regulating of cities are often created hierarchically. But it is important to see that order can also emerge from a spectrum of sources that extends from hierarchical and centralized types of authority to the completely decentralized and spontaneous interactions of individuals within a given community. Is there a place for the socially driven Black Rock City, where conduct and physical design are driven by common sense and common purpose, or are we destined to place our cities under the governance of prescriptive rules because we cannot reach a consensus on performance outcomes?

Social norms, particularly those that shape the design of places, should not be a rare cultural treasure unique to traditional societies, or temporary celebrations in remote deserts. Rather, they must be promoted and continue to flourish as a viable alternative to restrictive, one-size-fits-all scenarios. Indeed, social norms become even more important as technology becomes ubiquitous, economies globalize, and development is standardized. Such place-based norms are critical for neighborhoods to prosper and for development to be sustainable. For what is appropriate to be built and designed should be found not in the vision of an ideal average and social homogenization, but in the facts of cultural distinctiveness and in what is normal given the circumstances of place.

2

Experts of the Trade

The uppermost and undermost divisions of square furlongs, being measured from the external edges, or red lines, at the top and bottom of the town (as represented in this Plan), have a deduction of 41 feet 3 inches for half the width of the street on one side only: which, subtracted from 660 feet, the breadth of the furlong, leaves a space of 618 feet 9 inches for the breadth of 5 lots: Which space divided by 5 allows the breadth of each PLANTER'S TOWN-LOT in the uppermost and lowermost divisions to be 123 feet 9 inches including the fences; and the PLANTER'S town-lots, in the 4 central divisions (as the central divisions have a street on 2 sides), must lose 82 feet 6 inches from the width of each furlong; which being first deducted from the 660 feet leave a breadth of only 577 feet 6 inches to be divided into 5 equal parts, whereby the width of the PLANTERS lots in the 4 central divisions, is reduced to 115 feet 6 inches each, including the fences.

—Granville Sharp, 1794, From: *A General Plan for Laying Out Towns and Townships, on the New-Acquired Lands in the East Indies, America, or Elsewhere*

In shaping the contemporary form of urban Riyadh, Doxiadis Associates of Athens followed in the footsteps of many experts of the trade. For the last few centuries, urban standards have been shaped by the establishment of professional disciplines and their specific paradigms of practice. Most notable were the new professions that grew out of the traditional disciplines of law, divinity, and medicine. These urban form-shaping professions—the land surveyors and the civil engineers—in the absence of guilds had to adopt, endorse, and apply specific models to consolidate their positions and define themselves as experts. With their reserved body of knowledge, qualifying associations, and control through specific training and testing, these newly created specialists could restrict the role of outsiders to that of uninformed participants in no position to challenge the solutions of the "experts."

While the rise of this new breed of specialists can be attributed to the changing economic landscape of sixteenth- and seventeenth-century Europe, it was also tied to the growing trends of exploration and colonialism. Exploration and colonialism inspired many of the intellectual frameworks for the new social codes and the ordering of physical space. These frameworks often relied on central authority, favored grand plans and layouts (cities), and often disregarded the natural environment. Colonialism and imperialism also brought about methods and practices for land division and mapping that can be traced back to the techniques developed by the Romans. After conquering their neighbors, the Romans would inventory and record the wealth of their newly acquired colonies. Part of this process required a comprehensive survey of the landscape.

The Romans did not map these territories based on data derived from natural geographic features. Rather they implemented a cadastre plan—a conceptual grid of preestablished units of land division based on a coordinate system—and applied this plan to the actual land intended for subdivisional delineation. They surveyed the land according to the cadastre plan and marked the property boundaries. From this, two maps were created; one map was earmarked for the Roman archives, and the other, a map called a forma, remained with the colony. Property disputes within the colony were quickly and easily settled by referring to the forma.

The early Roman surveyors were called *agrimensores* (measurers of land). The *groma* was the primary tool of the *agrimensores'* art. This tool, a staff with four arms and four weighted string lines, allowed the surveyor to sight straight lines and determine right angles. Consequently, Roman land divisions were laid out in squares or rectangles. (See figure 2.1.)

The simplicity of this measuring and surveying system continued to appeal to future generations of surveyors and engineers. The urban historian Spiro Kostof traces this influence to the laying out of new cities and suburbs in the Renaissance (1400–1600), and Baroque (1600–1750) periods.[1] Regularly shaped spaces, uniform widths of streets, and standardized frontages can be found in many cities of these periods, with the two most regimented ones

Figure 2.1
The early Roman surveyors were called *agrimensores* (measurers of land), and the *groma* was their primary surveying tool. (*Source:* © Eran Ben-Joseph)

being Berlin and Dresden. In Dresden, for example, the process of issuing building permits included detailed planning and reviews based on a set of architectural standards. Applicants were obliged to submit both their own proposal and drawings showing their proposal in relation to the rest of the neighboring houses and outlying streets.[2]

Rationalized cities, controlled by strict form-shaping standards, were the physical manifestation of the Renaissance and Baroque rulers' affinity for regulating social behavior. Codes for designing streets and houses in St. Peterburg, Russia, for example, were part of Peter the Great's directives, which also mandated the shaving of beards and the wearing of particular dress.[3] According to Catherine the Great, such clean and constructed urban order made "the inhabitants . . . more docile and polite, less superstitious."[4] (See figure 2.2.)

Frameworks for the construction of orderly places were directly implemented in the colonial actions of the European powers. Unlike the sixteenth-century Laws of the Indies for New Spain, which prescribed the characteristics of towns but left their design to the settlers, other colonial powers such as England, France, and Portugal stipulated uniformity and order. In 1766, for example, Governor Luis Antonio de Souza of Portugal wrote that the new town of Santos should follow given standards, for "one of the things to which the

Figure 2.2
Rationalized cities, controlled by strict form-shaping standards, were the physical manifestations of governments' preference for regulating social behavior. Examples range from the buildup of grand, uniform streets and buildings such as Fredrichstrasse in Berlin and Dvortsovaya Square in St. Petersburg, Russia (photo). (*Source:* Courtesy © Alexey Sergeev)

most cultivated nations are giving their attention at the present time is the symmetry and harmony of the buildings of new cities and towns. This is not only of practical benefit, but also gives pleasure, in that such good order expresses the lawfulness and culture of the inhabitants."[5] (See figure 2.3.)

The Molders of Urban and Rural Landscapes

In the early nineteenth century the older established disciplines of law and medicine strived to set standards of competence and to define and enforce professional practices.[6] Newly established skills and trades, such as those provided by surveyors and engineers, imitated these precedents. For an occupation providing services rather than the production or selling of goods, asserting professionalism was of vital importance. As demands for their services grew, the character of the demands shifted from the aristocratic private sector to the governmental public one. Here, consistency, proficiency, and qualification had to be recognized.

In the early 1800s no professional standards or regulations governing the establishment of the surveying or engineering professions existed. Surveying

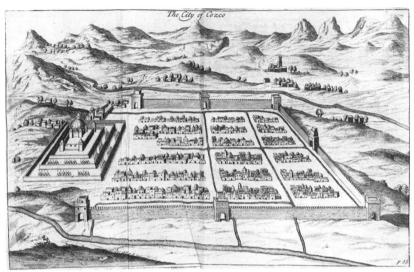

The City of Cozco

was a craft passed on from father to son, with reputation and seniority being the yardstick for measuring competence.[7] Overseas, the new lands of colonial conquest, however, demanded the competence of surveyors, while in the home country the industrial boom in canals, railroads, mines, roads, and towns and cities required ever-higher levels of professional performance.

In London, for example, fears of urban fires and harm associated with poor housing construction led to governmental building acts and the institu-tionalization of the overseeing professions. In 1844, the passage of the first Metropolitan Building Act created a new category of surveyor—the district surveyor. For the purpose of sharing experience, securing uniformity of prac-tice, and advancing sound principles of professionalism, aspiring or practicing surveyors had to pass an examination before their official appointment.[8]

The Building Act of 1844 can be considered one of the most compre-hensive regulations of the time. It evolved from earlier ordinances that con-trolled party wall thicknesses as well as building heights and materials, and was mainly instituted after London's Great Fire in 1666. Unlike these specific building guidelines, the 1844 act established town planning principles. Its preamble declared: "In many parts of the Metropolis and the Neighborhood thereof the Drainage of Houses is so imperfect as to endanger the Health of

Figure 2.4
This image of London in the nineteenth century shows the results of the 1707 Building Act requiring the construction of party walls between dwellings. Although compliance with the law ensured a greater measure of fire safety, it did not necessarily create better living conditions. (*Source:* © Giraudon/Art Resource, New York)

the Inhabitants, it is expedient to make Provision for facilitating and promoting the Improvement of such Drainage: And forasmuch as by reason of the Narrowness of Streets, Lanes and Alleys, and the Want of the Thoroughfare in many Places, the due Ventilation of crowded neighborhoods is often impeded, and the Health of the Inhabitants thereby endangered, and from close Contiguity of the opposite Houses the Risk of Accident by Fire is extended, it is expedient to make provision with regard to the streets and other Ways of the Metropolis for securing a sufficient Width thereof."[9] (See figure 2.4.)

Regulating street widths and building setbacks was seen as a key to controlling growth and promoting better living conditions by bringing light, air, cleanliness, and relief from congestion. Unfortunately the physical manifestation of these acts often resulted in wide, straight, and uniform streets and buildings out of place and insensitive to natural and social conditions. Sir Patrick Abercrombie, the English town planner, while acknowledging that these acts were "an enormous advance in many directions," wrote in 1933: "The Model By-

Laws prescribed the size of rooms, the space behind the house and the width of the road in front. If these things were satisfactory, each house supplied with air, light and access, what more could be wanted? The by-law period of town planning has surrounded all our business towns with rings and quarters of houses at an average density of 40 to 50 per acre, and has planed away every feature of natural interest and beauty. The only structure which these towns possess is given by the old country roads which penetrate through these blocks of by-law houses."[10]

The 1844 act also marked the beginning of professional regulation through minimal short diploma courses for those entering public service. The desire to protect their professional territory against the potential threat of architects and lawyers moving in on them motivated surveyors to create a professional association. In 1868 some twenty surveyors met at a London hotel to form the Institution of Surveyors and to outline the laws and regulations for the organization.[11] From their initial formation, it was realized that to establish their reputation and distinguish themselves as a profession they must require better education and ensuing examinations. In 1881 the institution was granted its Royal Charter on condition that examinations be introduced. With the requirement of such an exam, the surveyors in England established themselves as pioneers in the introduction of a professional qualifying examination, with the Royal Institute of British Architects and the civil engineers soon to follow.[12]

The Land, the Plan, and Gunter's Chain

The land surveyors were the colonial instruments of spatial order in the new territory. They were the appraisers, the delineators of routes and dwellings, and the shapers of urban and rural places. Until they traversed the land with a chain and compass, and recorded the results on a map, it could not be fully converted into developable property. (See figure 2.5.) As new lands were obtained, a first priority was to have them surveyed. In India, for example, surveyors gradually mapped the whole subcontinent. In South Africa and Rhodesia, surveyors used tight cadastral controls and a grid as "the quickest and simplest method of laying out the town."[13] In New South Wales and Western Australia, the surveyors' inspired layout of Adelaide became a model for other settlements in that colony and in New Zealand.[14] (See figure 2.6.)

The colonial powers, through their surveyors, often expressed a political authority and power through physical form. Organic settlements, such as those

Figure 2.5
Land surveyors were the shapers of colonial settlements and new lands. Through their actions, such as this marking of the land in Peru in 1730, appraisals, tax collection, and delineation of properties and routes came to fruition. (*Source:* Courtesy NOAA Central Library)

Figure 2.6
This plan of Adelaide, Australia, in 1837—laid on a grid pattern—provided a prototype for development in Australia and New Zealand. (*Source:* Courtesy Adelaide Public Library)

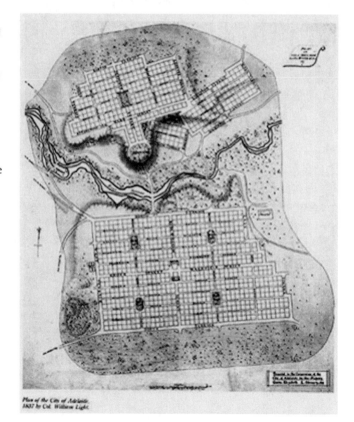

of the Islamic cities, were often replanned and divided by wide streets with the classical ideals of symmetry, order, and proportion. Such wide and straight streets not only organized cities into manageable units, but also fulfilled symbolic functions: they imposed a sense of order, facilitated police control, and were perceived as promoting healthy living places filled with light and fresh air.

Surveyors became change agents, and like the earlier *agrimensores,* they needed a tool and technique to transfer political ideals onto the land. Once land and the lines drawn on it were seen as a political manifestation of powers tied to economic wealth and value, knowing its exact size and configuration became essential. Units of measurements had to conform and they had to be accurate.

In the early 1600s, the English mathematician and inventor Edmund Gunter offered a solution. A professor of astronomy at Gresham College in London, Gunter was obsessed with ratios and proportions. He invented an early slide rule. In addition, he created tables of logarithmic sines and tangents and is believed to have been the first to introduce the words *cosine* and *cotangent* for sine and tangent of the complement. Gunter also realized that complicated mathematical and trigonometric calculations were beyond the ability of most surveyors and navigators. Therefore, like today's electronic inventors, he set out to translate his calculations into useful devices.

In 1606, Gunter suggested the use of a special chain to measure distances and areas. Unlike the rod and other chains used at the time, this chain was both convenient and accurate. Made of straight metal pieces, the chain did not stretch, shrink, or tear and was easily carried in a sack or on one's shoulder. Its length was composed of 100 links, marked into groups of 10 by brass rings. While at first glance, the dimensions of each link at just under 8 inches seemed to make no sense, in reality the device incorporated some of the most advanced numerical calculations of the period—the combination of the traditional English land-measurement system based on the number 4, and the newer decimal system. (See figure 2.7.)

In utilizing Gunter's chain, a surveyor could use either method and, if necessary, could easily convert dimensions from one system to the other. With 100 links, each at 7.92 inches long, the chain always equaled 66 feet. Since the traditional rod equaled 16.5 feet, a chain contained exactly 4 rods. A mile, at 5,280 feet long, contained exactly 80 chains. Most important for measuring

Figure 2.7
In 1606, Edmund Gunter suggested the use of a special chain to measure distances and areas. It became one of the most widely used instruments for measuring land across the globe. (*Source:* Courtesy Library of Congress Prints and Photographs Division)

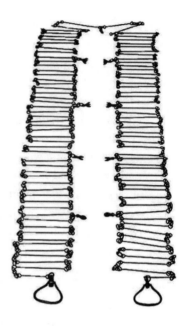

areas, one square chain (a square having sides that are each 1 chain, or 66 feet in length) contained 4,356 square feet, and since one acre contains 43,560 square feet, it could be measured by exactly 10 square chains.

In 1607 and afterward, it was hard to find another measuring system that was so consistent and convenient for calculating areas. Together with the *groma,* the straight line, the perpendicular angle, and the acre provided the means for making property a commodity. Once the layout of places became connected to business activities, simplification of urban form for economic benefits became the ideal standard. The fundamental unit of place making was no longer the neighborhood, but the individual lot, whose value could be gauged in length, width, and area. This rectangular system, with its precise measurements and organized boundaries, and its obliviousness to local conditions, has created a dimensional standard based on economic benefits rather than the creation of neighborhoods. As described by urbanist Lewis Mumford: "With a T-square and a triangle, finally, the municipal engineer could, without the slightest training as either an architect or a sociologist, 'plan' a metropolis, with its standard lots, its standard blocks, its standard street widths, in short, with its standardized, comparable, and replaceable parts."[15]

Disposing Land

The integration of this simple surveying technique with the desire to control land for political and economic benefits resulted in one of the most significant planning acts in the United States. On May 20, 1785, the Continental Congress adopted *An Ordinance for Ascertaining the Mode of Disposing of Lands in the Western Territory.* As the original Thirteen Colonies ceded their authority to the federal government after the Revolutionary War and Native Americans were forced to relinquished lands, government surveyors were instructed to divide the territory into individual townships using the chain. Each township was to be square, with each side of the square to measure 6 miles in length for a total of 36 square miles of territory. The town was then divided into 1-square-mile sections, with each section encompassing 640 acres. Each section received its own number, with section 16 to be set aside for a public school. The federal government reserved four sections to provide veterans of the American Revolution with land for their service, with the rest being sold at public auction.

While the dimensions of each square may seem arbitrary at first, they are utterly rational for someone using Gunter's chain. With each side of the square measuring 6 miles, or 480 chains, the number could easily be divided to the thirty-six smaller square-mile lots each containing 6,400 square chains. These could then be divided into halves by at least seven times to produce 5-acre units or 50 square chains, therefore allowing even the most unskilled surveyors to mark the land for sale or exploitation.

According to the U.S. Department of Agriculture, almost four-fifths of the U.S. land subdivision has been influenced by the systematic rectangular land survey.[16] Indeed, anyone flying over the United States can immediately see the imprint of this rectangular land-surveying system. Roads, streets, and fields are all impressed on the landscape in a perfect grid with straight lines, shunning any adjustments for natural conditions. In fact, these checkered patterns are so perfect that according to the U.S. National Aeronautics and Space Administration (NASA), their demarcation can even be distinguished by astronauts from space.[17]

The systematic rectangular survey system may have served a useful purpose for a nation driven by an agricultural economy and the desire to distribute land in a democratic yet orderly fashion. Yet its consequences for land-use patterns and urban form were probably never anticipated by the men who inscribed

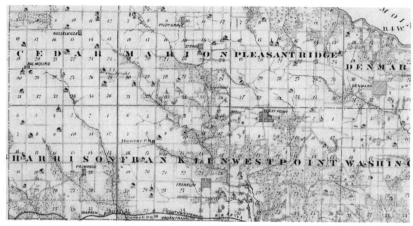

Figure 2.8
The Western Territory's
division into rectangular
townships and the disre-
gard of natural topogra-
phy can be clearly seen
in this partial map of
Iowa. (*Source:* Courtesy
Library of Congress
Prints and Photographs
Division)

it in 1785. The road network developed under this land subdivision, for example, completely disregarded natural features such as streams and rivers. Unlike the early settlers in Virginia and Massachusetts, who sited trails and roads
according to the lay of the land and minimized river crossings, most of the surveyors' roads along the systematic rectangular survey areas cross streams and
hills much more often and at a costlier rate.[18] (See figure 2.8.)

The straight lines of the 1785 ordinance, and the simple technology
offered by Gunter's chain, have also put their mark on city streets and town
planning. The experts involved in planning this new frontier could hardly resist the convenience and ease offered by the rectangularity of the system. The
original layout of Cleveland, Ohio, for example, was directly derived from
Gunter's chain and the straight lines of the required land subdivision. In 1796,
when Moses Cleveland, a surveyor for the Connecticut Land Company, found
the spot to build the capital for "New Connecticut" near a quiet bend of the
Cuyahoga River, he utilized the same technique he mastered while surveying
straight lines across the Western Reserve. The perfectly laid town grid with its
central square was configured by the chain. The largest road, Superior Street,
measured 2,640 feet in length and 132 feet across, or exactly 40 chains long by
2 chains wide. Minor streets were set at one and one-half chains in width, or
99 feet, flawlessly delineating the individual two-acre lots of 2 chains wide by
10 chains deep. All of this regular geometry encircled in turn the central 10-
acre plaza of 10 by 10 chains.[19] (See figure 2.9.)

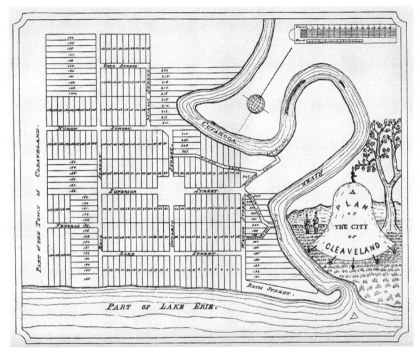

Figure 2.9
The original layout of
Cleveland, Ohio, was di-
rectly derived from
Gunter's chain. For ex-
ample, Superior Street
measured 2,640 feet in
length and 132 feet
across, or exactly 40
chains long by 2 chains
wide. (*Source:* Courtesy
Cleveland, Ohio, Histori-
cal Society)

As the land survey continued to establish its patterns across the West, new settlements, railroad towns as well as rebuilt old cities yielded to its systematized order. Railroad companies like Union Pacific designed standard towns that could be uniformly laid out on both sides of their tracks. The typical 160-acre sections on either side of the track were further divided into four 40-acre or 20-by-20 chain plots traversed by gridded streets. A letter to the *Atchison Daily Champion* on January 18, 1868, describes one such designed city, Waterville, Kansas, along the Central Branch Railroad:

I drop you a few lines from this, the newest and best town yet started by this enterprising company. Major Gunn was here yesterday, arranging for laying of the town into lots and blocks. It is a beautiful site, lying on a gentle slope from the bluffs toward the Little Blue River, and exactly 100 miles from Atchison, the acknowledged railroad centre of Kansas, and the company has acquired title to about 600 acres. . . . Already Mr.

Frahm has a boarding house started. . . . The company's head carpenter, the inevitable and irrepressible Bertram, is here with a large force of carpenters, finishing off the depot, 60 feet long, and will commence tomorrow upon the engine house, which is to be 60 feet in length. . . . I hazard nothing in saying that, in one year from this date, Waterville will be one of the neatest little towns on the line of this road.[20] (See figure 2.10.)

The power of the standardized rectangular system overshadowed renegades such as Washington, D.C. Indianapolis, which tried to emulate the diagonal avenues of the nation's capital, had to fit its radiating boulevards inside the survey section's 1-mile box. And San Francisco, which was offered a diagonal boulevard plan by architect Daniel H. Burnham, could have implemented it after the city was devastated by the 1906 earthquake and fire, but instead chose to reconstruct its previous grid.

The City-Street Shapers

Gridded rectangular cities such as San Francisco not only were a product of efficiency of surveying techniques but also had their roots in the effectiveness of military systems concerned with logistics and movement. The spread of these layouts in Europe was often related to Roman camp design and the soldiers who often were first inhabitants of these military installations turned towns. The English historian and archeologist, Francis John Haverfield, suggests that "when chess-board planning came into common use in the Roman Empire, many—perhaps most—of the towns to which it was applied were 'coloniae' manned by time-expired soldiers. So, too in the Middle Ages and even in comparatively modern times, the towns laid out with rectangular streetplans in Northern Italy, in Provence, in the Rhine Valley, are for the most part due in some way or another to military needs."[21]

Polybius (203–120 BC), the Greek historian of the Mediterranean world, provides us with one of the most detailed descriptions of the rigidity by which these Roman camps were planned. In book 6 of his *Histories,* he writes:

From the line described by the front of these tents they measure another distance of a hundred feet towards the front. At that distance another parallel straight line is drawn, and it is from this last that they begin arranging the quarters of the legions, which they do as follows—they bisect the last mentioned straight line, and from that point draw another straight line at right angles to it; along this line, on either side of it facing each other, the cavalry of the two legions are quartered with a space of fifty feet between them, which space is exactly bisected by the line last mentioned. . . . The result of these arrangements is that the whole camp thus forms a square, with streets and other constructions regularly planned like a town.[22]

With the grid came the value of the wide street. Traffic by horse, wagon, and later the automobile resulted in broad avenues as well as straight lines and grids traversing landscapes and crowded places. In 1879 George B. Selden, an American attorney, applied for a patent for a gasoline-powered, self-propelled vehicle—the automobile. Thirty-five years later, in 1914, annual automobile sales in the United States had reached more than 1.5 million. Six years later, in 1920, almost 10 million automobiles were traveling on an inadequate and poorly maintained road system. Changes in the construction and planning of transportation networks and streets were bound to happen, causing the reconfiguration of our surroundings.

Although state and local governments were desperately expanding their road networks, a coherent national road system had yet to be developed, one that would be coordinated financially and technically. This could only be achieved by federal action. In 1921 the Federal Highway Act was passed to provide for federal aid to construct "such projects as will expedite the completion of an adequate and connected system of highways, interstate in character."[23] The act was the first recognition in American transportation policy of the desirability of providing for a functional specialization for motor-vehicle routes and their control by a central authority. Such emphasis created the basis for a hierarchical road system and the first official categorization of roads and streets, in particular the separation of arterial through-traffic networks from local ones. Federal monetary aid had also generated the largest road-improvement program in the nation. During the Depression years in particular, federal aid was extended to include urban and rural road systems. By 1938, road and street improvements reached a total of 600,000 miles, as compared with only 80,000 miles for highway construction.[24]

A New Profession

The change in policy and the improvement of road systems necessitated the emergence of a new profession. Except among a few engineers, there was little available knowledge of the fundamental differences between road-construction techniques and transportation planning. Many of these early professionals were civil or mechanical engineers, self-taught in transportation planning and construction.

In the late seventeenth and early eighteenth centuries most American civil engineers worked either on canal or railroad projects. Those who received any formal training were typically graduates of West Point. However, many practicing engineers were self-taught or trained on the job.[25] Attempting to distinguish themselves from these low-ranking subordinates, high-ranking engineers aspired to follow in the footsteps of their English counterparts and incorporate into a professional organization. They saw in such a formal national institute, a vehicle for giving its members prestige and professional standing that would encourage companies (particularly rail companies) to single them out for jobs. While local societies of civil engineers formed in Boston in 1840 and New York in 1852, it was only in 1867 that the national American Society of Civil Engineers was established. Similarly, the establishment of land-grant colleges in the 1860s offered a venue for professional engineering education far removed from the military academies and the common on-the-job training.

The need for specialization in road and traffic engineering as a result of rapidly changing transportation requirements impelled the formation of the transportation engineering profession in 1930 through the national Institute of Transportation Engineers (ITE) and a specialized education program at Yale University. The new profession was to be "a branch of engineering which is devoted to the study and improvement of the traffic performance of road networks and terminals. Its purpose is to achieve efficient, free, and rapid flow of traffic; yet, at the same time, to prevent traffic accidents and casualties. Its procedures are based on scientific and engineering disciplines. Its methods include regulation and control, on one hand, and planning and geometric design, on the other."[26] In 1939, the ITE was approached for the first time by the federal government, the National Conservation Bureau, and the American Association of Highway Officials to suggest traffic-engineering guidelines and standards in the form of an engineering handbook and related technical publications. In

Figure 2.10
Railroad companies like Union Pacific designed standard towns such as Waterville, Kansas, to be uniformly laid out on both sides of their tracks. The typical 160-acre sections on each side were further divided into four 40-acre or 20-by-20 chain plots traversed by gridded streets. (*Source:* Courtesy Kansas State Historical Society)

1942 the *Traffic Engineering Handbook* was first published and provided the basis for the profession and its practice.

Standardizing the Residential Street

In the 1940s, traffic engineers recommended lane widths and intersections that emphasized driver comfort, safety at high speeds, and the overall efficient movement of vehicles. Specific issues associated with the safe passage of through traffic in neighborhoods were not addressed by traffic engineers until the mid-1950s. At that time the engineers hoped to stop through traffic in residential neighborhoods by means of a hierarchical street network that would funnel through traffic to the main arterial roads. However, this approach did not change the inherent geometric configuration of the streets on which such traffic flowed. Moreover, the majority within the traffic engineering profession remained more concerned with building efficient, high-speed road networks, rather than with designing streets that reflected and sustained valuable aspects of a local neighborhood's character.

The ITE later turned its attention to neighborhoods. In the 1960s it published its *Recommended Practices for Subdivision Streets.* The text stated that "the primary objective of subdivision design is to provide maximum livability. This requires a safe and efficient access and circulation system, connecting homes, schools, playgrounds, shops and other subdivision activities for both pedestrians and vehicles." After articulating such principles, the publication proposed standards aimed at creating efficient vehicular movement. Thus, the document as a whole presents a conflicting message: first proffering flexible principles for street layouts and then subsequently overruling itself with a set of rigid standards.

On the one hand it stated: "Although it is extremely important that sound standards be followed in the layout and design of neighborhoods and of neighborhood street systems, it is equally important that there be room for variety, experimentation and improvements in residential design." On the other hand its charts and measurements spelled out such detailed requirements as a minimum right-of-way of 60 feet, pavement widths of 32 to 34 feet, and cul-de-sacs with a maximum length of 1,000 feet and a 100-foot diameter at their end.[27]

The ITE standards have been widely used as a basis for regulation by local agencies and public works departments ever since. Because they emanate from and are endorsed by a professional source, considered by default to be an ultimate and indisputable authority, such standards are assumed to be accurate, scientific, and based on empirical research. At a minimum, beyond the question of their appropriateness for all situations, these standards are depended on and defended as the solutions to the problems of designing critical infrastructure. They provide the added benefit of shielding local authorities from the personal liability that may be the accidental result of any traffic-layout innovation or creativity. With the residential street standards sanctioned by the professionals, new standards (or even modifications) have been very slow to develop. Since high-level governmental agencies have not felt driven to advocate changes, local agencies are reluctant, lacking any compelling reasons of their own, to do so either.

A critical component is missing from this perpetuating influence of decades-old standards. The role of the transportation profession and the process through which plans are initiated, developed, and approved often exclude a social perspective. Working together to the exclusion of broader perspectives, engineering and public works departments have long taken the lead in guiding the development and management of streets. Their practice has reinforced a prevailing emphasis on traffic performance with limited (if any) concern for practical accessibility and livability. The frequently cited failure of streets to provide adequate accessibility due to an increase in mobility and car ownership has resulted in design standards that often exceed traffic requirements and negate the social and livable component of neighborhood design. Without a social perspective, residential streets have become nothing more than traffic conduits, a means of getting vehicles from one point to another. (See figure 2.11.)

Residential streets should be more than traffic channels. Primarily, they should be the place for community interaction and neighborhood development. Streets, particularly in residential areas, provide a visual setting, a meeting place for neighbors, a play area, an entryway to and from each house, and a pedestrian circulation system. Ideally, their design requires an understanding not just of traffic patterns and capabilities, but of multiple users, social behavior, architecture, and urban design. However, these "fuzzier" elements are often hard to measure, and more to the point, do not fit within the professional paradigm of engineering practice. These limitations in perspective have even been acknowledged by reflective engineers, who admit that "the engineer is a person who is trained to create good and, in some cases, new solutions to well-defined problems. However, when he is confronted with problems which are complex or not well defined, he gets into difficulties. . . . If for example there is a strong interrelation between variables that cannot be described in a simple quantitative way, he tries to ignore this interrelation or he has the tendency to consider variable elements in a first approximation as constants. This is exactly what has happened in the field of geometric design standards for roads."[28]

Typical standards for street layouts, like those issued by the ITE, are specification standards. That is, they are codes that specify the features of objects (so many feet or units of) or the range within which features must lie (i.e., at least so many feet of, or between x and y units of). Performance standards, similar

Figure 2.11
Efficiency in subdividing was at the core of engineering and planning practices. Shown are examples of recommended practices that included the following warning: "The legitimate subdivisions are impaired by the poor ones, because the latter glut the market and depress the entire building development field. Some authoritative observers believe that such 'wildcat' subdivisions have constituted not only a contributory factor, but a major primary factor in the economic collapse of the 1930's. Here is a vast blunder in social and economic policy, a blunder which will be extremely difficult to rectify." (*Source:* American Society of Civil Engineers)

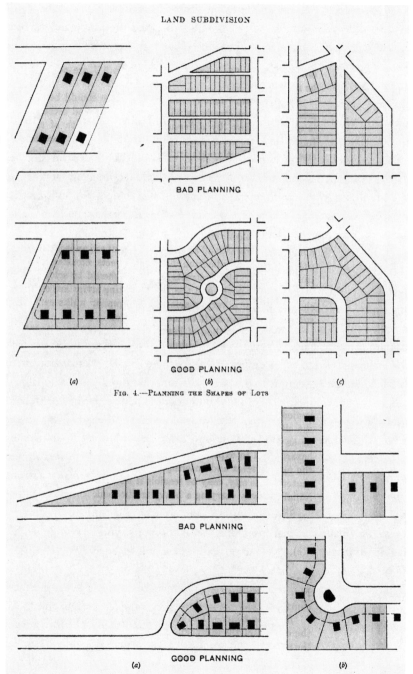

LAND SUBDIVISION

BAD PLANNING

GOOD PLANNING

(a)   (b)   (c)

FIG. 4.—PLANNING THE SHAPES OF LOTS

BAD PLANNING

GOOD PLANNING

(a)   (b)

to the norms applied in the Islamic city, do not specify how things must be, but what they must or must not do or what their capacity or impact must be. Rather than work by means of restrictions, this form of regulation shapes the built environment by imposing limits on the impacts of that change, leaving greater flexibility in design and construction. It allows for a mixture of different types of solutions on a given site, where traditional specification standards call for conformity.

Reestablishing norms and using performance standards in the design of places is not an easy undertaking. For the last century and a half we have shifted our planning apparatus from indicating and recommending values to specifying and requiring explicit standards. The attraction to planning by specification can be understood in relation to the establishment of new professions such as those of surveyors and engineers in the seventeenth and eighteenth centuries, as well as in relation to the deteriorating conditions of nineteenth-century cities. For example, between 1870 and 1900 the population of the United States doubled, and by 1910 almost every other American lived in an urban area compared to slightly more than one in four in 1880. Congestion, overcrowding, and worsening sanitary conditions were believed to cause social and moral degeneration. Social reformers argued that social disorder, which was bound to happen, would be best disciplined by proper environmental conditions. As demand for better conditions in the city grew, pressure increased on local governments to restrain socially harmful activities, to provide new services, and to take steps toward controlling the physical shape of the city.

Against the chaotic environments of congestion and social unrest of the late nineteenth century stood the ideal of a disciplined technological utopia with a compelling spatial order. The ideal found its realization in a new direction for civic improvement that emerged to reform the environment and control it through the employment of expert knowledge, state regulatory mechanisms, and public-welfare provisions.

3

Neighborhoods Developed Scientifically

City planning is not a fad today, it is a necessity; it is not an extravagance, it is an economy; it is not an artist's dream, it is a scientific reality.

—Arnold W. Brunner, 1912

The World's Columbian Exposition in Chicago in 1893 was an image of a new utopia. The fair's vision of immense, white neoclassical buildings facing naturalistic landscapes and sparkling lagoons, stood in bright contrast to the dark reality of typical city living. Even after its closing and demolition (six months after opening day), its image remained as a powerful reminder and inspiration for advocates of reform. They argued that city improvements would create civic patriotism and a better disposition toward the community, and that beautiful surroundings would enhance worker productivity and urban economies. Although the main aim was to improve aesthetics through public buildings, civic centers, parks, and boulevard systems, advocates of this City Beautiful Movement also included commonplace residential improvements, particularly to tenements and slums.

The precedents of English surveys and European social statistical studies led Congress to authorize an 1892 investigation of the conditions in American slums. By 1900, more than three thousand surveys had been produced, many of them by private organizations.[1] With the prevailing spirit of technology and science, rational planning and utilitarian ethics were summoned to guide public policy. (See figure 3.1.) Rationality inspired the adoption of the German concept of zoning and street systems, and it led to the adoption of the English comprehensive plan. Such pressures for professional solutions to cities' chaotic environments prompted the First National Conference on City Planning and

Figure 3.1
At the end of the nineteenth century, congestion, overcrowding, and unsanitary conditions fostered a popular concern for public health. Tenements and slums were the first focus of many early planning remedies, which attempted to bring light, air, and healthful living through a scientific frame of reference. (*Sources:* Courtesy © CORBIS (a) and Library of Congress Farm Security Administration (b))

the Problems of Congestion held in Washington in 1909. This conference was the first formal expression of widespread professional interest in a systematic approach to solving the problems of America's urban environment. Congestion stood at the forefront. The remedies proposed by the conference sought to encourage private enterprise to reduce center-city crowding. It was believed that the population of cities could be redistributed by relocating the middle classes to the suburbs. As a result, these classes would cease to compete with the lower classes allowing the lower classes to obtain better housing. Reformers hoped the older housing would serve as a base for upward social mobility, while home ownership in new suburbs would establish social and economic stability there. The provision of fast and low-cost transportation and the location of industry at the fringe would together facilitate population redistribution.

The 1909 conference attracted the attention of senators and representatives, and President William Howard Taft showed his interest by consenting to make the opening address. This conference, and those that followed, laid the groundwork for city planning structure and formed its implementation techniques. Issues, such as "The Best Methods of Land Subdivision" and "Street Widths and Their Subdivision," established land planning standards based on contemporary public-health and housing concerns. They also created the foundation on which the federal, state, and local governments erected zoning and subdivision regulations a few years later.

In this conference, and those that followed, participants struggled with the notion of control despite the lusty and unruly growth of cities. The nation's belief in the sanctity of land speculation, land ownership, and property rights seemed to stand in opposition to controlling and regulating. In a speech titled "The Control of Municipal Development by the 'Zone System' and Its Application in the United States," Antrim Haldeman, the Assistant Engineer for the City of Philadelphia, remarked: "The necessity for limiting the right of the individual to do as he pleases has arisen from the exploitation of the property and rights of the public by private interests, and from the exigencies attending the intensive growth of great cities."[2] Others continued the call for centrally coordinated intervention and the scientific approach to solving the city's ills. In 1912, for example, Frederic C. Howe, one of the founders of the American city planning movement, declared:

Our cities are what they are because we have not thought of the city as a city, of the town as a town, of the rights of everybody as opposed to the rights of anybody. A million men are thinking only of their individual lot lines, of their inviolable right to do as they will with their own, irrespective of its effect on the community. We do not see beyond our own doorsteps, we do not think in city terms, or appreciate that the progress of society has so far socialized old conditions that the community must have a life of its own separate from, or the composite of, the lives and property of all of its people. We have exalted the rights of the individual above the common weal. Our cities have been permitted to grow with no concern for the future and with no thought of the community or the terrible costs which this uncontrolled development creates.

This failure to think in community terms, to appreciate that the city is a physical thing involves costs which the future cannot repair. And the most costly blunder of all is our neglect of the city's foundations, of the land on which the city is built. The American city is inconvenient, dirty, lacking in charm and beauty because the individual landowner has been permitted to plan it, to build, to do as he willed with his land. There has been no community control, no sense of the public as opposed to private rights.

Our cities have been planned by a hundred different land owners, each desirous of securing the quickest possible speculative returns from the sale of his property. Streets have been laid out without regard to the needs of the future. They have been cheaply paved, watered, and sewered. There have been few building restrictions, little provision for parks, open spaces or sites for public buildings.[3]

If the unregulated ways of city building proved so costly in community resources and the inhabitants' well-being, then science and technology could offer socially and economically superior results. By regulating the cities' physical living conditions, social ills could be relieved. In these years, public officials, including President Theodore Roosevelt, stated that "the conservation of our national resources is only preliminary to the larger question of national efficiency."[4] Efficiency, in turn, could only come about through the replacement of speculating owners with experts and technocrats who could enforce rational policies and administer scientific solutions. (See figure 3.2.)

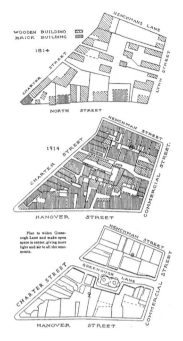

Figure 3.2
Typical survey plan of excessive development densities over time, and the proposed remedy through street widening. North End, Boston. (*Source:* Massachusetts Homestead Commission)

City Practical

Frederick Winslow Taylor, the most influential expert of that period regarding management practices, and the author of *The Principles of Scientific Management* (1911), believed that the goal of human labor and thought is efficiency. He saw technical calculation as superior to human judgment. In fact, he believed that humans could not be trusted because they are plagued by laxity, ambiguity, and unnecessary complexity. He measured by the universal units of science, not by the specifics of craft, person, or place. For Taylor, that which could not be measured either did not exist or had no value. He wrote:

> In the past, the prevailing idea has been well expressed in the saying that "Captains of industry are born, not made"; and the theory has been that if one could get the right man, methods could be safely left to him. In the future it will be appreciated that our leaders must be trained right as well as born right, and that no great man can (with the old system of

personal management) hope to compete with a number of ordinary men who have been properly organized so as efficiently to cooperate. . . . This involves the gradual substitution of science for rule of thumb throughout the mechanic arts. . . . The development of a science, on the other hand, involves the establishment of many rules, laws, and formulae which replace the judgment of the individual workman and which can be effectively used only after having been systematically recorded, indexed, etc. The practical use of scientific data also calls for a room in which to keep the books, records, etc., and a desk for the planner to work at. Thus all of the planning which under the old system was done by the workman, as a result of his personal experience, must of necessity under the new system be done by the management in accordance with the laws of the science; because even if the workman was well suited to the development and use of scientific data, it would be physically impossible for him to work at his machine and at a desk at the same time. It is also clear that in most cases one type of man is needed to plan ahead and an entirely different type to execute the work.[5]

The principle of scientific management captured the minds of industry leaders because it seemed to ensure calculated conduct that would lead to increased profits. Its attitudes and orientations soon spread to the movement for city planning. In an article published in 1913 in the popular magazine *The American City* and titled "Efficiency in City Planning," the authors proclaimed that "the principles of modern industrial efficiency, of 'Taylorizing' are now being applied to city planning."[6] They stated that "this method of work, systematized, standardized, 'Taylorized,' as it is, has most decidedly proved its worth. It applies strongly to the businessman, the man who has to pay the bills, and convinces everyone that the experts have real knowledge on which to base their recommendations, and are not presenting mere dreams, pretty but impracticable."[7]

Such statements expressed the attitude of many experts and town planners who at the time attempted to shift the consideration of aesthetics and the notion of City Beautiful to the act of city planning. George Ford, a prolific urban planning consultant and in later years the director of the New York Regional Plan, spoke at the fifth annual meeting of the National Conference on City Planning in 1913 about "The Scientific City." Ford announced that "city planning is rapidly becoming as definite a science as pure engineering."[8] He advocated planning actions based on scientific procedures—data collection and analysis by experts—as well as the resulting systematic solutions. "By such cooperation and by standardized procedure," he wrote, "it is possible to determine within a comparatively short time a plan which is not only the best for to-day but which is so elastic that any changes during the next fifty or one hundred years can be fitted into it with virtually no loss or alteration."[9]

The notion of the practical city, built according to scientific, standardized principles, was seconded by many designers. In 1917, the publication of the American Institute of Architects, *City Planning Progress,* stated: "City planning in America has been retarded because the first emphasis has been given to the 'City Beautiful' instead of the 'City Practical.' We insist with vigor that all city planning should start on the foundation of economic practicableness and good business; that it must be something which will appeal to the businessman, and to the manufacturer, as sane and reasonable."[10]

In their search for a scientific model for the American city, these newly formed experts looked overseas. According to Fredrick Law Olmsted Jr., the president of the National Planning Conference and the first president of the American City Planning Institute (1917), the model was found in Prussia (Germany). After the Franco-Prussian War (1870–1871), many Prussian cities experienced an uncontrolled buildup of inner-city factories and housing, followed by dangerous health hazards and pollution. Citizens protested the deteriorating environmental conditions. Prussian city officials responded with a zone system that assigned particular land uses to designated locations. The Prussian zone system also called forth new principles of design. In trying to avoid the evils of American land speculation, German planners rejected the American grid system and chose instead a hierarchical, varied street system to order the zones. As described by Howe, who studied the German zoning system for the American Planning Conference in 1911: "Germany turned her trained intelligence to the control of the physical side of the city; to the control of property, as we control persons whose license is inimical to the community. Private property was subordinated to humanity, while the speculator, builder, and factory owner were required to use their own as the community decreed."[11]

One of the influential physical planning laws of that era was the 1875 Prussian Law Concerning the Laying Out of and Alteration of Streets and

Squares in Cities and Country Places. The law provided the legal basis for numerous municipal regulations, particularly the layout of new, wide streets to foster healthful conditions. Many sanitary and urban reformers in Europe and the United States advanced this proposition for laying out wide streets that allowed sun and air to reach the crowded blocks of houses. Josef Stübben, an Assistant Burgomaster (Mayor) and Royal Counselor of Buildings in Cologne, was a Berlin-trained architect who was involved in city planning studies of more than thirty cities in Germany and abroad. As one of Europe's best-known planners, Stübben advocated what he called the "Practical and Aesthetic Principles for the Laying Out of Cities."[12] He argued against the organic unplanned form of the medieval city and pushed for regulations of the physical form of the city in order to create healthy places. In his influential handbook for city planning, *Handbuch des Städtebaue,* the first edition of which was published in 1890 with revised and expanded versions following in 1907 and 1924, Stübben established the specific dimensions for "the systematic or regulated laying out of cities."

These practical principles included the regulation of street layout and design, the forming of blocks and appropriate subdivisions, and the allotment of sanitary and healthy environs. He wrote:

> The essential basis of regulated city-building is the establishment of future streets and places, and, consequently, of the plans for them. Here we regard not so much the technical improvement of the streets, or street-building proper, but rather a very skillful, consistent connection and complete network of streets, requisite for the traffic and growth of the city, the residence and joint life of the citizens. We are constrained, therefore, to develop the mere street plan into a building plan, or, further, into a city plan, i.e., a design in which, out of the network of streets and building areas, are arranged all those constructive regulations as well, which are demanded for the life of the city.
>
> The grades of the streets are to be made as convenient as possible for the traffic: the maximum to be according to local circumstances from 1 to 10%; the minimum, for the sake of drainage, 0.5 to 0.2 per cent. It is desirable that the streets should lie slightly above the general surface (say, 1 m.). The width of streets should correspond to the expected traffic. It should be at least 10 m.; more than 40 m. is required only in the

rarest instances. As to wider streets the claims of health and beauty give the proportions.

Transversely, the street is to be divided into at least one roadway and two raised sidewalks. A certain amount and kind of travel requires a special walk or promenade in the center of the street, a special bridle path or the division of the roadway for teams and equipages, giving rise to various cross-sections, and so much the more, when, for reasons of health and beauty, the streets are set out with rows of trees and flower beds.[13]

This notion of directing city development through design codes and regulations was widely admired in Britain, the United States, and Scandinavia. In England, the well-established Public Health Act, aimed at improving the health and sanitation conditions of the British towns, granted code and regulatory powers to local authorities. As with the German law, these included stipulations for the physical design of the city such as the arrangements and configurations of streets. The bylaw street right-of-way of 60 to 70 feet rested on the justification of access to light and air. It also set the basic configurations of street widths that remain the cornerstones of residential street standards used today.[14]

The German notion of zoning and physical improvement through regulations seemed to be a model for American cities. Howe writes again:

House-building can be controlled, easily controlled. German cities limit the amount of land that can be built upon in the business sections to from 65 to 75 per-cent; in new sections to 35 per-cent of the lot areas. They limit the height of buildings, usually to the width of the street. They provide that sunlight shall have a chance to enter into every story. It is this community control that gives the German city its charm.

Factories are required to build in those sections away from the direction of the prevailing winds, so that the smoke and dirt will be driven away from the city. Cities open up parks near factories so that working-men may have a convenient place to rest and play. Germany controls its factories in the interest of human life and efficiency.[15]

Although German zoning addressed problems of securing decent environments for both city centers and their growing edges, in the United States the adoption

Figure 3.3
Plan showing existing
conditions and a pro-
posed layout for a
"blighted area to be im-
proved," in the South
End section of Boston.
With the introduction of
zoning techniques in the
United States, existing
mixed-use urban areas
were gradually trans-
formed into homoge-
neous districts, and
newly designed places
were deliberately
planned for uniform use.
(*Source:* Boston City
Planning Board)

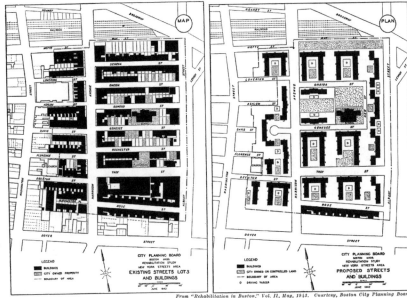

From "Rehabilitation in Boston," Vol. II, May, 1943. Courtesy, Boston City Planning Board.

of the German precedent began with the concerns of the center city, and only
later moved outward to the suburban edge, where it took on a new direction.

Because in the United States the control of private land is a state, county,
or municipal function, zoning began as an intensely local practice. Each jurisdiction made its own zoning map, usually beginning by setting aside very large
areas for high-value industrial and commercial uses. In general, in the early
years the industrial zones lay along the railroad rights-of-way, and the commercial and apartment-house zones followed the streetcar lines. Residential
neighborhoods, whatever their inherited mix of uses, were grandfathered in as
they stood. Out on the unbuilt edges of town planners established the future
residential neighborhood zones. (See figure 3.3.)

Residential Development

In the early days of urban development and expansion, subdividing and establishing neighborhoods was basically regulated through surveying rules, methods, and practices. The aim was to provide a more efficient technique for selling

land, permitting the recording of a plat of land by dividing it into blocks and lots that were laid out and sequentially numbered. The platting facilitated the sale of land and prevented conflicting deeds. Uniformity was seen as a way to facilitate both surveying methods and the assessment of property.

Land speculation, uncontrolled growth, and inadequate building construction in the late nineteenth and early twentieth centuries raised many concerns about the acts of subdividing the land. Premature subdivision created an oversupply, leading to the instability, and ultimate deflation, of property values. (See figure 3.4.) Depreciation of economic value led to tax delinquencies and widespread foreclosures. Partial development of tracts often resulted in conflicting property titles, misaligned streets, increased costs, and reduced provisions for public amenities. President Herbert Hoover, an engineer and a firm

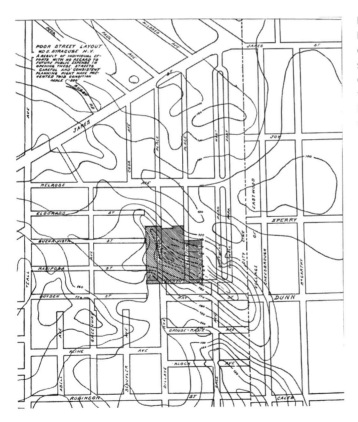

Figure 3.4
Patchwork subdivisions of tangled property lines and broken street alignments resulted in parallel movements for building codes, street surveying, and, ultimately, twentieth-century use and structure zoning as well as subdivision controls.
(*Source:* City of Syracuse)

believer in local control, sponsored the publication in 1928 of *The Standard City Planning Enabling Act* (SCPEA). In addition to serving as a tool for recording and conveying property, and setting forth models for local zoning, it emphasized the need for on-site improvements to support the demands created by the new subdivisions. Road layouts, block sizes and lots, sidewalks, and drainage facilities were addressed as ways to ensure minimum standards of construction, livability, and control of development itself.

In these early years of zoning regulation a number of experts, including city planners, started to pay more attention to efficiently controlling and designing new subdivisions. Thomas Adams, an Englishman specializing in residential developments, who became one of the founding members of the American City Planning Institute in 1917, was particularly concerned with inefficient and wasteful subdivision practices. In his book *Neighborhoods of Small Homes,* he wrote: "In connection with both the problem of building new houses and improving old houses there is a need of more knowledge of underlying economic conditions. This is particularly so in regard to the cost of land development and the necessity or otherwise of the unhealthful densities of buildings that are allowed to prevail in large cities."[16] Adams, together with Robert Whitten, the president of the American City Planning Institute, embarked on an economic study to compare and analyze different planning schemes. Their analysis concentrated on ratios of costs associated with different physical factors such as densities, open space, utilities, street widths, and landscaping. In their conclusion and recommendations, they see the self-contained neighborhood superblock as the best and most economical solution.

By the early 1930s, Adams had attained such a prominent standing as an expert in subdivision development that many of his writings were adopted as recommended practices by government and professional institutions. In 1931, President Hoover's Conference on Home Building and Home Ownership used suggestions Adams made in his forthcoming book *The Design of Residential Areas* as part of its recommended practices. The president's conference was the largest ever to be held by the federal government. More than 3,700 experts on home finance, taxation, and planning of residential districts formed committees and put forward a multitude of recommendations that shaped the built landscape for generations to come.

The call for such an extensive conference was rooted in the uncontrolled practices of subdivision planning that resulted in unruly platting and chaotic marketing. The call was made urgent by the harsh realities of the economic depression that had turned American municipal authorities into ineffectual entities. As the Committee on Planning for Residential Districts described: "Too much current practice in municipal development is based upon habit, insufficient vision, excessive speculation in land, and over emphasis of new growth upon the outskirts to the detriment of older sections, which too often become blighted. . . . Individual action and individual decisions on matters of concern to the whole community have been the rule on community development. Shanty-towns, houses off grade and askew with the street, unsanitary conditions, and unsightly developments have resulted. Suburban slum areas have been created. Subdivision practices and the contribution of municipalities to subdivision development are in need of review and of regulation."[17]

The conference's Subdivision Layout Subcommittee came to the conference determined to set new standards and regulations. In its adaptation of Adams's treatise, it called for the reduction of subdivision costs through "thoughtful planning and functional layout."[18] Through Adams's comparative diagrams and charts, it endorsed the neighborhood as the principle unit for organizing place, and the interior cul-de-sac as the most economical configuration for platting.

Four years later, in 1936, to further encourage coordinated local planning, the Advisory Committee on City Planning and Zoning—appointed by the Secretary of Commerce—published the Model Subdivision Regulations. By 1941 thirty-two states had passed legislation granting the power of subdivision control to counties and municipalities through the establishment of local planning commissions. Through an exercise of legislative "police power" by the state, the right of a landowner to sell property could be withheld until approval by a designated authority that was mandated to "promote the community health, safety, morals, and general welfare."[19]

Such a law can be seen in the following Commonwealth of Massachusetts example, which stated in part:

Subdivision control law has been enacted for the purpose of protecting the safety, convenience and welfare of the inhabitants of the cities and towns . . . by regulating the laying out and construction of ways in subdivisions providing access to the several lots therein, but which have not become public ways, and ensuring sanitary conditions in subdivisions and in proper cases parks and open areas. The powers of a planning board . . . under the subdivision control law shall be exercised with due regard for the provision of adequate access to all lots in a subdivision by ways that will be safe and convenient for travel; for lessening congestion in such ways and in the adjacent public ways; for reducing danger to life and limb in the operation of motor vehicles; for securing safety in the case of fire, flood, panic and other emergencies; for ensuring compliance with the applicable zoning ordinances or bylaws; for securing adequate provisions for water, sewerage, drainage, underground utility services, fire, police, and other similar municipal equipment, and street lighting and other requirements where necessary in a subdivision; and for coordinating the ways in a subdivision with each other and with public ways in the city or town in which it is located and with the ways in neighboring subdivisions.[20]

The Massachusetts Act, and similar legislation across the nation, empowered local boards to regulate subdivision practices within their jurisdictions. It was enabling legislation, not a set of standards, so that it allowed each community to judge for itself how a particular subdivision proposal might fit the local circumstances. It promised a good deal of variety, even experimentation, but variety and innovation did not turn out to be the American path. In an unforeseen way, uniform standards came to dominate subdivision controls everywhere.

Neighborhood–Design Standards

At the time when communities established subdivision controls, many localities searched for a blueprint of technical regulations. What should be the character of a specific neighborhood and how does one nurture it? What is the best configuration for lots? Where should neighborhood amenities be placed? Should its streets be laid out in a gridiron pattern or in picturesque curvilinear fashion? How wide should those streets be?

Some of the answers were provided through the ideal of a constructed neighborhood and the community center movement.

The concept of a neighborhood as the unit for planning and design appealed to many experts at the turn of the twentieth century. Based on the goals of social reformers and on utopian visions of community-oriented yet integrated hygienic cities, the neighborhood was envisioned as the way to improve residential conditions in cities plagued by slums and uncontrolled speculative growth. Period writings such as Ebenezer Howard's 1898 *Tomorrow a Real Path to Real Reform,* Jacob Riis's 1902 *The Battle with the Slum,* and Charles Cooley's 1909 *Social Organization* were instrumental in inspiring designers and planners to translate conceptual ideas into physical form. Howard's vision of "a town designed for healthy living and industry; of a size that makes possible a full measure of social life, but not larger; surrounded by a rural belt; the whole of the land being in public ownership or held in trust for the community," materialized in Raymond Unwin and Barry Parker's design of Letchworth and Hampstead Garden Cities in England, and later on in Clarence Stein and Henry Wright's plans for Sunnyside Gardens, New York, and Radburn, New Jersey.[21]

Riis's ideals of city districts utilizing public schools as centers for social and civic pride had diverse roots. Social workers and educators had long sought to link schools to neighborhoods and the school curriculum to the needs and interests of individual children. Public officials and politicians saw schools as an agency of Americanization in a pluralistic society. Some social reformers and educators like Jacob Riis and Mary Parker Follett had regarded the public school as a nucleus in the most direct sense—a physical community center in the heart of a neighborhood.

This unofficial trend enjoyed great advocacy and political backing. Slowly the vision spread from the neighborhoods of New York City to the rest of the nation. In 1911 the first National Conference on Civic and Neighborhood Center Development was held in Madison, Wisconsin. In its conclusion the conference endorsed schools as social centers and agencies of reform.[22]

Neighborhood Scientific

Seeking to translate the social goals of the community center movement into physical action, a few cities sought to reorganize their existing and planned neighborhoods. Two such examples are the City Plan for St. Louis in 1907, and the Chicago City Club's national competition of 1912.

The City Plan for St. Louis provides one of the first links between the idea of neighborhoods as social entities and a physical plan for a city. Calling for neighborhoods to be planned around a civic core or a civic building such as a school or police station, the authors of the plan write:

> [A neighborhood center] would give a splendid opportunity for an harmonious architectural and landscape treatment of the various buildings, thus adding to the intrinsic beauty of each; it would foster civic pride in the neighborhood and would form a model improvement work, the influence of which would extend to every home in the district; it would give to the immigrant—ignorant of our customs and institutions—a personal contact with the higher functions which the government exercises towards him, as manifested in the only municipal institution with which he is brought in contact—the police station—a feeling that the government is, after all, maintained for his individual well-being as well as that of the native-born inhabitant. Lastly, it would develop a neighborhood feeling, which in these days of specialization has grown weak, with a resulting lack of interest in local politics and the consequent corruption and disregard of the best interests of the people by their representatives.[23]

While these social ideals envisaged a physical neighborhood prototype, they did not provide principles or design standards for their creation. It is in Chicago's progressive social and political setting that we find the impetus for translating theory into action.

In 1913 the City Club of Chicago held a national competition "addressed particularly to building and landscape architects, engineers and sociologists, for competition for plans for laying out, as a resident district, a typical area in the outskirts of the city."[24] Almost all entrants produced plans of residential developments laid around a central civic building. Variations between the plans were mainly taken in the design of streets and street layouts, the distribution of open spaces, and the architecture of the buildings. Almost all entries suggest and discuss the ability of their plan to be replicated and copied across the city along adjacent quarters. (See figure 3.5.)

PLAN SHOWING POSSIBLE REPETITION OF QUARTER-SECTION UNIT

Architect William Drummond even went as far as to coin a term for his design element: the *Neighborhood Unit Plan.* As he puts it,

> Order is the keynote of our plan. It provides that the whole city be divided into areas approximately such as the quarter-section. Each of these areas is regarded as a unit in the social and political structure of the city.
>
> Quarter-section may prove to be too large or too small, but the unit is intended to compromise an area which will permanently exist as a neighborhood or primary social circle. Each unit has its intellectual, recreational, and civic requirements featured in the institute which is located approximately at its center and its local business requirements featured at its corners.[25]

At the same time in the Chicago School of Sociology, students and faculty were actively involved in studying the city as a natural human environment while emphasizing the importance of human communities as shapers of the city.

These sociologists highlighted social research and social reform, and were also concerned with the expanding metropolis and the influence it exerted over adjoining neighborhoods. One of the doctoral students in the school, Roderick D. McKenzie, completed his dissertation on a neighborhood study of Columbus, Ohio. McKenzie took an ecological approach from the study of plant science and applied it to human communities. He noted the influence of geography, space, and the environment on human social organization and change. The research led him to draw an analogy between nature, where centers of dominance develop in higher organisms that coordinate and control the organism, and similar phenomena occurring in human communities. And as these societies and communities evolve they become "more differentiated and the points of dominance more concentrated."[26]

In 1923 *The Neighborhood: A Study of Local Life in the City of Columbus, Ohio* was published. In this book McKenzie expanded on his thesis to include an outline of historical information about the neighborhood as a locus of community organization from Western antiquity to modern times.

Neighborhoods for the Machine Age

While McKenzie's book and the idea of a neighborhood as a design element grew in popularity in sociological circles, it still lacked both architectural exposure and simplified design parameters that could be applied to any place or locale.

Clarence Perry, a social worker who worked for the Division of Recreation at the Russell Sage Foundation on Community Issues, and who was active in the community center movement, showed great interest in setting universal parameters for applying the neighborhood unit idea. Perry felt that despite wide acceptance of after-hours use of schools, the influence on the adjacent neighborhoods was minimal. He realized that better integration between the surrounding neighborhoods and the center must occur and that all elements of the urban plan must be dependent on each other. His objective remained the same—the creation of an institutional basis for socialization and cooperative citizenship on the neighborhood level—but the means shifted to a standardized plan for the physical environment to achieve it.[27]

Although he had developed the essentials of his prototypical plan as early as 1923, the concept had its full impact on the design of residential developments in 1929, with the publication of the *Regional Plan of New York and Its Environs,* in which Perry published a section on "The Neighborhood Unit: A Scheme of Arrangement for the Family Community."

Perry, who directly borrowed Drummond's neighborhood-unit terminology, provided in this influential publication six standards for a successful neighborhood unit "which embraces all the public facilities and conditions required for the average family for its comfort and proper development within the vicinity of its dwellings."[28]

Size   related to an elementary school and encompassing 750 to 1,500 families on a 150 to 300 net acre site within a quarter-mile radius extending out from the school to encourage walking

Boundary on All Sides   Marked with arterial streets with 120 foot right-of-way to eliminate through traffic in the neighborhood

Internal Street System   designed to discourage through traffic with 50 foot rights-of-way

Open Space System   of small parks and recreation spaces

Institutional Site   for a school and other neighborhood institutions grouped at the center; and

Local Shops   placed at the edges of the unit, at traffic junctions, and adjacent to other neighborhoods.[29]

Perry's work contributed to the acceptance of the residential neighborhood as a special entity that needed to be protected and deliberately planned for. (See figure 3.6.) Although innovative and extensively promoted, the concept was not quickly adopted. In his 1939 book *Housing for the Machine Age,* Perry himself acknowledged that the chief obstacle to effective implementation of the concept was the prevailing emphasis on small-scale building enterprises in the United States and the lack of comprehensive planning policies able to implement projects on a large scale.[30] It was through the eventual work of Thomas Adams, the endorsement by President Hoover's 1931 Conference on Home Building and Home Ownership, and the backing by the federal government and professional organizations that the concept began to take shape on the ground.

Neighborhoods Developed Scientifically

Robert Whitten, City Planner, has made a brilliant contribution to the science and art of City Planning by his development of a Neighborhood Unit—a tract of 160 acres, with a population of 5000 to 6000, with no through traffic, with no increment in land values—earned or unearned—where homes for workers can be built for the equivalent of a rental of $50. a month—where home and neighborhood with schools, parks, playgrounds and stores will be *permanent.*

ROBERT WHITTEN
City Planner, N. Y.

The Federal Housing Administration's endorsement of the neighborhood unit and its standard-setting publications had a lasting effect on suburban development. (See figure 3.7.) In 1936, for example, the FHA's Technical Bulletin No. 5—*Planning Neighborhoods for Small Houses*—demonstrated the Administration's preference for the neighborhood planning concepts. Using plans and diagrams as well as language that appeared in Perry's publications, the bulletin states:

> The importance of distinctive neighborhood qualities lies not only in the initial appeal which is so vital a factor in marketing the development, nor in the increased security which derives from the safeguards created by careful planning, but also in the psychological reaction of the people who adopt the area for their homes. Where a neighborhood can be identified and comprehended as such, the feeling of pride and reasonability which the owner has in his own parcel, tends to be extended to the neighborhood as a whole. A sense of community responsibility and a community spirit thus develops, which acts as a stabilizing and sustaining influence in the maintenance of realty values.
>
> This larger interest cannot develop where no neighborhood identity exists. Developments which consist of no more than additional blocks in

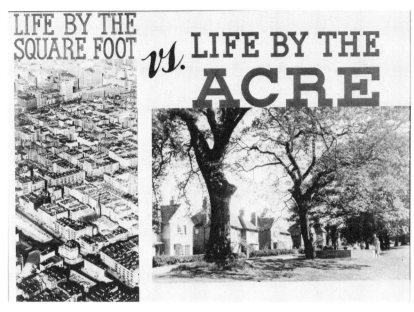

Figure 3.7
Neighborhood-unit de-
velopments and subur-
ban resettlement were
advocated by profes-
sional associations and
government agencies.
Posters such as this one
by the U.S. Resettlement
Administration empha-
sized the dichotomy be-
tween the congested
unplanned cities and the
designed, lush new
neighborhoods at the ur-
ban edge. (*Source:* Li-
brary of Congress Farm
Security Administration)

an already monotonous accumulation of blocks, or properties which are strung out as ribbon developments along main thoroughfares cannot provide this greatest of all safeguards. Subdivisions of these latter types can bring neither a sustained market to the developer nor stable values to the purchaser and are, therefore to be discouraged.[31]

Between 1940 and the late 1970s, the neighborhood-unit concept was used extensively in new developments in North America and in the development of many new towns around the world.[32] The scope of this standardized approach to addressing the physical and social ills of modern cities can also be seen in its endorsement by health organizations, particularly the American Public Health Association (APHA).

The American Public Health Association: Planning the Neighborhood

On April 18, 1872, a group of doctors met in a New York hotel to form a national group that became the American Public Health Association. Inspired by international and national interest in civic sanitation, and by attempts to pass a national health-board law that would create local health boards, these doctors sought to address the problem of contagious diseases through uniform practices, health education, and a national health policy.

In its early years the association exerted pressure on the government to control the deteriorating health conditions in American cities. One of its most ambitious projects was a survey of sanitary conditions across the nation, which yielded reports on water supplies, drainage and sewage, streets and public grounds, housing, garbage, slaughterhouses, and health laws.[33] In the late 1800s the APHA began making recommendations for standardizing water analysis, culminating in the publication of *Standard Methods for the Examination of Water and Sewage* in 1895, a precursor to a book still in print today.

Standardization became an important element in the APHA's campaign, and its standardized manuals became landmark codes of practice. *Standard Methods for the Examination of Milk* was published in 1905, followed by *Standard Methods for the Examination of Air* in 1909. Far reaching and influential, these documents and those that followed—such as *Control of Communicable Diseases in Man*—were used internationally and translated into other languages, including Arabic, French, Italian, and Chinese.[34]

With the growing endorsements of the neighborhood-unit concept by urban planners and social reformers, health officials realized its potential benefits. In 1948 the association's Committee on Hygiene of Housing published the document *Planning the Neighborhood*. The goal was to set minimum standards to encourage developers to build neighborhoods "that improve the quality of life, for all incomes" while maintaining cleanliness, providing adequate daylight, sunshine, and ventilation, protecting against excessive noise and pollution, and allowing opportunities for community life as well as guarding against moral hazards.[35]

The document goes on to establish the neighborhood as the basis for its standards, claiming that a "unified neighborhood positively influences stability and development of family life." Building on Perry's publications and those by the FHA, it codes in great detail various design parameters, from how to deal with various hazards to the optimal size of streets, playgrounds, and parks. The document further warns about the location of "Moral Hazards" in relation to planned neighborhoods and states that "the site and surrounding area should be checked for the existence of undesirable establishments including gambling houses, bars, low-grade taverns and nightclubs, and houses of prostitution.

Such establishments in the vicinity, especially along paths leading to schools and other facilities with heavy use, should be avoided."[36]

As for the overall physical layout of the neighborhood, the association proposes prescriptive dimensions to be followed; these contain many of Perry's standards as well as other dimensions influenced by various FHA publications. They include:

Neighborhood boundaries and size These features are defined by the limit on how many people can be served around a school, with 5,000 inhabitants as the most desirable size for a neighborhood population. Schools should be within a half-mile walking distance and ideally within a quarter-mile walking distance.

Neighborhood density For one- and two-family detached houses, density should not exceed seven units per net acre; for semidetached houses of this type, it should not exceed twelve units per acre. To provide adequate sunlight, building coverage for multiple dwellings should be limited to 40 percent.

Circulation-system design Streets must be classified and designed according to their use. They should include residential service streets with low traffic; neighborhood feeder streets with moderate interior traffic; minor traffic streets that connect feeder streets to major traffic streets and generally act as boundaries of the neighborhood; and major traffic streets that connect cities and districts, with long-distance, high-speed traffic.

Sidewalks should always be provided on at least one side of the street with a minimum width of 3 feet, and up to about 5 feet for streets with major pedestrian flow.

Outdoor recreation and open space Three to four acres of parks should be provided per 1,000 persons. The minimum park size should be 1.5 acres connected to the residential areas and community facilities with well-lighted paths.

Neighborhood playgrounds of 2.75 acres—serving 120 children—should be allocated, with maximum size not to exceed 6 acres. The playground must be surrounded by fencing or other barriers, and the maximum finished grade is 2 percent.

The APHA's "Planning the Neighborhood," as well as "The Neighborhood Unit" of the *Regional Plan of New York,* were the products of professional associations whose members saw it as their duty to protect the health and welfare of residents. Originating in the desire to better the dreadful conditions of dense urban areas at the end of the nineteenth and early twentieth centuries, they signified the institutionalization of planning. In this fight for progress, standards became the essential tool for solving the problems of health, safety, and morality. Assuming the controls over neighborhood patterns and form, these standards offered sets of codes that could be part of a place-making toolbox.

Federal Intervention

While professional associations sowed the seeds of specifications and development standards, it was through the establishment of the Federal Housing Administration (FHA) that the government ended up shaping residential development in the most significant ways.

The FHA was established in 1934 under the National Housing Act. Its main goal was to restructure the collapsed private home-financing system through government mortgage insurance plans. Through the FHA's long-term, low interest rates and low down payments, a growing proportion of the population found itself able to buy a home and maintain affordable payments. At the same time, the nature of the government's plan also shielded lenders from financial risk. Developers in turn found new stimuli for existing project sales as well as incentives for new construction. Together, the FHA's financial assistance and mortgage insurance created the most ambitious suburbanization plan in American history.

The FHA's underwriting procedure soon turned into the prevailing standard for residential mortgages. To secure its investments, the FHA established a comprehensive system of appraisal procedures designed to eliminate risk and failure. To qualify for a loan, lenders, borrowers, and developers had to submit detailed plans and documentation to establish the feasibility and soundness of their proposed projects. With monetary support at stake, developers and local communities actively worked to comply with the federal standards.

It is important to note that the FHA's process was voluntary, not compulsory. Borrowers could of course seek financing elsewhere, but federal guaranteed loans carried the lowest interest rates and required the smallest down payment. And to qualify for federal loans, borrowers had to comply with the federal standards. FHA officials soon found themselves in a powerful posi-

tion—ironically, one of far greater influence than any planning agency—to direct and shape development for generations to come.

In 1934, nearly 4,000 financial institutions, representing more than 70
percent of all the commercial banking resources of the country, had FHA insurance plans. By 1959, FHA mortgage insurance had helped to provide homes
for 5 million families and to repair or improve 22 million properties. Three out
of every five American families were helped by the federal government to pur-
chase a home.[37]

Community developers and the National Association of Real Estate
Boards were enthusiastic about the FHA's intervention role, in contrast to their
fear and anxiety about the influence of government and local planning commissions. The FHA's successful control over development and developers was
not only due to its financial power, but also to the fact that it was *not* a real planning agency. Unlike planning agencies, the FHA was largely run by representatives from the real estate and banking industries, and thus developers felt their
needs were served. The FHA's dominant standards and underwriting supported large, established and prominent builders, who were able to further
expand and construct large-scale residential subdivisions with government
backing, and to put the 1920s speculative-style quick profiteering "Jerry-
builders" out of business. Thus, the paradox of the FHA system was that although it imposed strict requirements through underwriting manuals and
property standards, "it always appeared to be non-coercive to the private sector.
The FHA was generally perceived as engaging in a simple business operation—
to insure only low-risk mortgages with a sound economic future. Property
owners and real estate entrepreneurs viewed FHA rules and regulations as similar to deed restrictions—private contracts which were freely entered into by
willing parties—rather than as similar to zoning laws, which were sometimes
seen as infringing on constitutional liberties."[38]

In January 1935, the FHA's first technical-standards publication, *Standards for the Insurance of Mortgages on Properties Located in Undeveloped Subdivisions—Title II of the National Housing Act,* was published. Its companion,
Subdivision Development (Circular No. 5), was the basis for subsequent publications by the agency's Technical Division. In the latter publication, the FHA
stated its goals and visions for successful development in general terms. Attempting in its language to mute the perception of implied rules or obligatory

procedures, the document declared that "the Administration does not propose
to regulate subdividing throughout the country, nor to set up stereotype patterns of land development," yet it goes on to state (as if it couldn't help itself):
"It does, however, insist upon the observance of rational principles of development in those areas in which insured mortgages are desired, principles which
have been proved by experience and which apply with equal force to neighborhoods for wage earners as they do to those for the higher income groups."[39]
These general principles are then described in a detailed narrative, with precise
measurements given in the section "Minimum Requirements & Desirable
Standards." (See figure 3.8.)

Beyond setting the framework for regulation through written standards,
the FHA also provided suggestions and recommendations for residential-
development layout. The 1936 bulletin on *Planning Neighborhoods for Small
Houses* illustrates the measurements needed to build an ideal, "well-balanced,
carefully planned subdivision," one that would facilitate "the creation of real
estate values through devising a layout which is not only economically sound

Figure 3.8
The U.S. Federal Housing Administration's 1938
neighborhood-standards suggestions for dos and
don'ts for lot configurations and building setbacks. By the 1950s, the FHA's *Neighborhood
Standards*—published in the *Land Planning Bulletin No. 3*—were widely utilized in each state.
(*Source:* U.S. Federal Housing Administration,
Technical Bulletin No. 7)

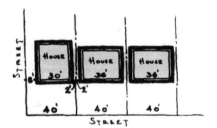

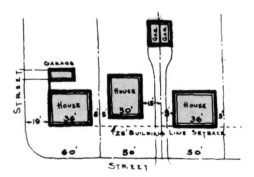

but which provides to the maximum degree those conditions which make for pleasant and healthful living."[40]

In its preferred standards for overall subdivision layout, the FHA rejected the grid pattern for residential neighborhoods; this approach is carried through all of its later publications and suggested requirements. The bulletin spelled out the FHA's objections to the grid pattern as follows: "The gridiron plan which has been so universally adopted in most of our cities has several very decided disadvantages when applied to residential areas. In the first place, it creates waste by providing a greater paved area than is necessarily adequate to serve a residential community. Secondly, it causes the installation of a more expensive type of paving by dispersing the traffic equally through the area, which in turn creates an increased traffic hazard. In addition to these disadvantages, it creates a monotonous uninteresting architectural effect and fails to create a community aspect."[41]

To illustrate its preferences, the FHA proposed three forms of residential street layouts: curvilinear, cul-de-sacs, and courts. Their designs were in each case defined by descriptive and prescribed measurements.

In 1938, the FHA initiated a free review program for prospective developers, to enable them to have their preliminary plans examined. The review procedure and required forms were included in most of the Administration's publications. Those forms typically stated that "the FHA is interested in cooperating with real estate developers, builders, and their technical consultants in obtaining high standards of land development. The opportunity is welcomed to analyze proposed subdivisions and to make suggestions which, in our opinion, if followed, will create more marketable, attractive, and stable residential properties."[42] FHA consultants would then analyze the plans submitted and, as needed, suggest the types of physical layouts that would conform to the FHA guidelines for securing an insured mortgage. It was a powerful control tool. Almost all subdivision developers submitted their plans for review, accepting the government's "preferences" as the reasonable cost of a ticket for safe passage to a guaranteed federally funded loan.

By such steps the federal government was able to exercise tremendous authority and power through the simple act of making an offer that could not be realistically refused. The extent of such authority was stated by the FHA administrator, James Moffett, in 1935, when in a confidential meeting he told his advisory board to "make it conditional that these mortgages must be insured under the Housing Act, and through that we could control over-building insertions, which would undermine values, or through political pull, building in isolated spots, where it was not a good investment. You could also control the population trend, the neighborhood standards, and material and everything else through the President."[43]

In these ways, the FHA's minimum standards and design regulations lay the groundwork for modern subdivision practice. They were the source for local government subdivision regulations, and in retrospect, there is clear evidence that they were the historical basis for the uniformity of the predominant residential design across the land. Indeed, this became the case even as local planning administrations were stating things like: "Good subdivision design cannot be standardized and applied universally to all tracts, but only basic principles and minimum standards of design can be formulated."[44]

This may have been thought to be true (then and now), but the evidence is clear: incorporating diversity and adapting development to local conditions have remained primarily rhetorical exercises, because most localities continue to regularly adopt a familiar, nationally prevalent set of recommended standards as a continuing place-making default.

Part II

Locked in Place

4

Sanitized Cities

> We can discover a difficulty in rightly judging the works of a city in the fact that innovations or changes are frequently resisted by those in charge either from a force of habit adhering to old customs or from the inconvenience of altering existing laws.
>
> —Rudolph Hering, 1881 (in *Report of the Results of an Examination Made in 1880 of Sewerage Works in Europe*)

Design standards and regulations are not limited to controlling and shaping streets, lots, and rights-of-way. As early as 1967, the Urban Land Institute warned that "the basic parameters for sanitary sewer design were set at the turn of the century and, for the most part, have remained unquestioned since that time. Sewerage collection systems today are designed almost by rote, picking values off charts and conforming to standards which were in existence before the present generation of engineers were born."[1] In the book *Land Use and the Pipe*, Richard Tabors, Michael Shapiro, and Peter Rogers further suggest that planners in particular feel inadequate in challenging proposals put forward to them because of a perceived lack of expertise and a general attitude of not being able to address engineering criteria and parameters.[2] (See figure 4.1.) The lack of public interest in and knowledge about sewage systems has both perpetuated old approaches and served officials and their technicians well. This view is acknowledged by the International City Managers Association: "One hears often of local situations in which the people reject the financing of a new city hall or a new central plaza in favor of an expenditure for improved sewage systems. It is rather fortunate in this connection that the people ask only for results, for this gives a freedom of action to the administrator and the technician."[3]

Figure 4.1
Since the 1800s engineers and public officials have preferred sewer removal and treatment solutions based on wasteful technologies. All use water as the carrying agent for the waste and therefore require thousands of treatment and collection facilities and millions of miles of pipes and conduits. (*Source:* Courtesy © Cross County Metro, St. Louis)

Indeed, in the next 20 years, American communities will spend over $1 trillion to upgrade and build new wastewater infrastructure. While federal assistance programs will provide 15 percent of the funding for metropolitan-area wastewater-treatment projects, it is the local communities that must provide the major share of funding for these projects.[4] While infrastructure shortfall can be attributed to various factors, such as aging pipes and inadequate treatment facilities, the questions to be asked are: Will the upgraded systems be any better? Will they be more appropriate to a changing world where groundwater shortage looms, freshwater resources are declining, and sustainability is vital? Or will the upgraded systems rely on entrenched practices and technologies that are outdated and inappropriate, rather than seeking new alternatives? (See figure 4.2.)

For over a century engineers have chosen sewer removal and treatment solutions from a small range of wasteful technologies. All are based on water as the carrying agent for the waste and are dependent on thousands of treatment and collection facilities and millions of miles of pipes and conduits. Poet Wendell Berry vividly describes this illogical situation as follows: "If I urinated and defecated into a pitcher of drinking water and then proceeded to quench my thirst from the pitcher, I would undoubtedly be considered crazy. If I invented

Figure 4.2
The Deer Island sewage
treatment plant is part
of a $3.8 billion program
to treat wastewater
from Boston-area com-
munities. It is estimated
that in the coming two
decades towns and cities
across America will
spend over $1 trillion to
upgrade and build new
wastewater infrastruc-
ture based on similar
technology. Other parts
of the world are follow-
ing the same paradigm.
In China, for example,
over 2,000 medium to
large treatment plants
based on typical waste-
removal technology will
be needed in the near
future. (*Source:* Courtesy
© Massachusetts Water
Resources Authority)

an expensive technology to put my urine and feces into my drinking water, and then invented another expensive (and undependable) technology to make the same water fit to drink, I might be thought even crazier. It is not inconceivable that some psychiatrist would ask me knowingly why I wanted to mess up my drinking water in the first place."[5]

Consider, for example, that in households not yet utilizing water-efficient fixtures, toilets use the largest fraction of the family's consumption—20 gallons per person per day.[6] This figure means that one person contaminates about 7,300 gallons of freshwater to move a mere 165 gallons of body waste each year. It also means that we destroy a valuable fertilizer resource by mixing it with water, making it just about useless.

Even more troubling are indications that such practices are being adopted and encouraged in countries that have extensive traditions of recycling human waste. China, for example, has a long tradition of ecologically sound wastewater recycling in multilevel biological systems based on aquaculture. Yet, during the current development process, and with the encouragement of international experts who are advocating state-of-the-art solutions, these practices have been stigmatized as old fashioned. Villages and towns are being encouraged to abandon them. Instead of the former methods, over 2,000 medium-to-large treatment plants based on typical waste-removal technology will be constructed in the near future.[7]

Do these systems represent the best solution available? Or are these engineering practices, and the planning apparatus that supports them, operating within an outdated technological paradigm? They seem to represent the anomaly of societies held captive to an existing technology in a world marked by voluntary decisions and pluralistic processes.

The paradigm of sewer infrastructure shows that dependencies on past decisions prevent the application of alternative technologies. Historical decisions about the methods and systems for sewerage collection have locked our current practice into a specific mode of operation. Such a situation means that ecologically appropriate alternatives are rarely advanced as options before decision makers and therefore cannot gain wide acceptance.

Urban Sewage

Although many ancient cities such as Babylon and Jerusalem had drainage channels to carry rainwater and waste, it was the Romans who systematically planned and constructed underground sewers to drain uplands to the nearby network of low-lying streams. Roman sewers developed from open channels that drained most of the land prior to urbanization. The philosophy was to use the existing natural drainage channels to remove stormwater. The city then was built over the channels, and drains were provided from the surface to the underground streams. Yet with the growth of large cities, particularly Rome, sanitary conditions deteriorated. Residents constructed numerous altars and shrines dedicated to Febris, the Goddess of Fever, and Verminus, the God of Disease, hoping and praying for improvements.[8]

While the prayers may have been answered by the gods and goddesses, they were also considered by the Roman engineers. By providing a general water supply and an elaborate drainage/sewer system, and by regulating burial grounds, Rome was gradually transformed into a less dangerous place. The Cloaca Maxima, for example, was an engineering marvel of the time. A 4.2-meter-high and 3.2-meter-wide vaulted, paved tunnel, it provided the main excess-stormwater disposal system for the city, as well as a sewage-collection system from the buildings along its route to the Tiber River. (See figure 4.3.)

Figure 4.3
The Cloaca Maxima was
an engineering marvel in
its time. A 4.2-meter-
high and 3.2-meter-wide
vaulted, paved tunnel, it
was the main excess-
stormwater disposal sys-
tem for Rome, as well as
a sewage-collection sys-
tem for the buildings
along its route to the
Tiber River. (*Source:*
Courtesy Arizona Water
& Pollution Control Asso-
ciation)

In the centuries after the decline of the Roman Empire, urban sanitary engineering practically vanished. The urban environment was filthy and unhealthy. When underground channels were constructed, they were designed to carry stormwater rather than human and household wastes. The impure water supplies and accumulated waste of medieval cities fostered terrible epidemics that decimated much of Europe. (See figure 4.4.) With the renewed interest in the study of science in the fourteenth and fifteenth centuries, attention again was paid to sanitary engineering. Still, surface and street drains, rather than sewers, were at the center of this development. Human waste was typically accumulated in cesspools or privies, and as late as 1815 an English law forbade

Figure 4.4
Until the early twentieth century, health officials believed in the ability of water to purify waste. Detail from a painting by Pieter Brueghel the Younger, 1559. (*Source:* Courtesy Staatliche Museen zu Berlin, Gemäldegalerie)

emptying waste into the street drainage system. With minimal application of available knowledge of hydraulics or topographical science, efforts to improve drainage or sewage systems were often ineffective.[9] (See figure 4.5.)

London typified the mid-nineteenth century experience of Europe. By the 1840s, London's population numbered over 2 million, living in several hundred thousand households. An awareness of the need for sewer-system reform and development led to the first comprehensive study of the metropolis for the purpose of planning improvements. In 1847, the first official report on sewer and drainage systems—by the engineer John Phillips—contained the following description, which portrayed a typical situation at the time:

> There are hundreds, I may say thousands, of houses in this metropolis which have no drainage whatever, and the greater part of them have stinking, overflowing cesspools, and there are also hundreds of streets, courts and alleys that have no sewers; and how the drainage and filth are cleaned away and how the miserable inhabitants live in such places it is hard to tell. In pursuance of my duties, I have visited very many places where filth was lying scattered about the rooms, vaults, cellars, areas, and yards, so thick and so deep that it was hardly possible to move for it. I have also seen in such places human beings living and sleeping in sunk rooms with filth from overflowing cesspools exuding through and run-

ning down the walls and over the floors. . . . The effects of the effluvia, stench and poisonous gases constantly evolving from these foul accumulations were apparent in the haggard, wan and swarthy countenances and enfeebled limbs of the poor creatures whom I found residing over and amongst these dens of pollution and wretchedness.[10]

The conditions in London resembled those of most cities of 200 years ago. By that time nearly every residence had a cesspool that collected and stored all

Figure 4.5
Fires and explosions due to methane buildup in unventilated cesspools under buildings were a familiar event in nineteenth-century London. (*Source:* Courtesy © Hulton-Deutsch Collection/CORBIS)

house waste beneath its first floor. With cesspool overflows, the failure of proper drainage, and the contamination of drinking water, epidemics and lingering illnesses became common. Fires and explosions due to methane buildup in unventilated cesspools were just as frequent. Such conditions are vividly described in the following account from 1849, when workers entering to examine cesspools with oil lamps triggered sudden blasts: "Explosions occurred in two separate locations where the men had the skin peeled off their faces and their hair singed. In advancing toward Southampton, the deposit deepens to 2 feet 9 inches, leaving only 1 foot 11 inches of space in the sewer. At about 400 feet from the entrance, the first lamp went out and, 100 feet further on, the second lamp created an explosion and burnt the hair and face of the person holding it."[11]

Facing these and other deteriorating urban conditions, some local authorities—usually known as "Improvement" commissions—sprang into existence. Through a series of Building Acts, these local authorities were endowed with limited powers to regulate the design, quality, and location of buildings and the management of waste. London instituted such an act in 1774, and Liverpool in 1825.[12] However, constitutionally, financially, and technically these commissions were ill-equipped to cope with the immensity of the task, and their impact was hardly felt. State intervention was needed if change was to occur. Sanitary reformers of the time, particularly Edwin Chadwick, argued for the establishment and extension of both local and central government authority. Such views are also reflected in both the 1842 *Report on the Sanitary Conditions of the Labouring Population of Great Britain* and the 1844 *First Report of the Commissioners of the State of Large Towns and Population Districts,* which advocated a fundamental rethinking of the regulatory framework. A new centralized mechanism was seen as the key to controlling growth and ensuring long-term planning. One of the reports' ultimate outcomes was the Public Health Act of 1848, by which, for the first time, the British government charged itself with the responsibility of safeguarding the health and well-being of its population through the establishment of the General Board of Health. The act fell short of establishing a comprehensive nationally managed sewer system, and left the design and planning to the local public-health boards. It did, however, legislate that no new housing was to be built in an area within the health boards' jurisdiction without suitable provision of sewer disposal. As stated in the General Board of Health Minutes: "The removal of all cesspools from amidst habitations is the first duty of local boards. As soon as a proper survey and system levels have been obtained, the first duty of a Local Board of Health will be the prosecution of measures requisite for the entire abolition of all cesspools, and the prevention of their further formation, by the complete drainage of every house in the towns."[13]

Standardizing Sewer Systems

The General Board of Health was also engaged in an active debate about the nature and form of the appropriate technology of waste disposal. This debate came about because of a new technology introduced to the cities at that time. This new technology was that of piped-in potable water and the flushing toilet.

As piped-in water became available, households rapidly installed water fixtures. The introduction of large volumes of wastewater into the cesspool system caused serious flooding, disposal, and health problems. In Paris in 1832, 20,000 people died of cholera. In other parts of Europe the combination of piped water and open sewers consistently led to similar outbreaks. Facing this crisis, the officials and experts argued over the following two options:

- Retention of the dry or semidry conveyance method for human waste while disposing other gray water through the existing stormwater system
- Combining human waste and gray water into one system to carry it toward natural water bodies

Advocates for the dry or semidry system argued that such a system would retain the valuable nutrients and economic benefits associated with dry decomposing waste. The ability to use the waste as fertilizer was already a common practice in Europe, as well as in many other parts of the world. The *Sydney Morning Herald,* for example, warned its readers in Australia in 1851 that "we shall not always be able to rob the soil, and give it nothing in return."[14] In an article under the heading "Utilization of Sewage," the *New York Times* wrote that "one of the greatest sources of waste in all American and most foreign cities, is the throwing away of the valuable fertilizing elements in the sewage.

In China, that semi-barbaric, but exceedingly wise land, the inhabitants of cities are obliged by law and custom to return all fecal matter to the country, where it may be restored to the soil, from which it came in the shape of vegetable and animal food. . . . The refructification of soils is getting to be an important question along the Atlantic States."[15]

Other experts, although acknowledging the economic benefits associated with recycling human waste, quickly (and in retrospect, wrongly) argued that waste in water has as many benefits as when it is disposed of in the earth. In 1881, while giving a lecture on sanitation, the Boston engineer Edward Philbrick said:

> The sewage of cities has often been used with success as a fertilizer for the soil. It has accordingly been argued by many that it must be a great waste for London, and on smaller scale for Boston, too, to throw their sewage into the sea. But experience has proved that the value of sewage as a manure is not so great as was once proposed, while the cost of applying it to the soil is often too great to be thus recouped. If thrown to the sea, the inhabitants of the particular district where this occurs may not profit by its value, but this is not lost to the world. It is soon decomposed by exposure and the elements of which it once consisted are stored in the air or water for the future use of either vegetable or animal life. The kelp is nourished and contributes to human wants, so do the various mollusks that fatten upon the mud in our bays.[16]

In an article titled "Sewage Carried by Water," *Engineering Magazine* refers to studies done by Chadwick and remarks: "Mr. Chadwick, in his report on the Paris Exhibition on dwellings for the working classes, has drawn attention to a number of most important sanitary questions, of which the above is one of them, and at page 76 of this able report points out, that fresh sewage that has not undergone the process of decomposition is not only more valuable in its fructifying power in the proportion of one to three; but, instead of killing fish when it escapes fresh into the river, the fish come and feed on it."[17]

Such arguments continued on both sides of the Atlantic. In a series of articles titled "Modern Sewer Construction and Sewage Disposal," published in *The Sanitary Engineer and Construction Record,* engineer Edward Philbrick comments on Boston's 1875 Mayor's Commission on the Sanitation Conditions. He writes:

> The most specious arguments against the scheme of disposal came from those who believed in the value of sewage as manure and advised its application to the land where such value could be availed of. It was called a sinful waste to consign so much organic matter to the ocean, and thereby ignore the advice of those celebrated modern chemists who had taught us its use as plant-food. Theoretically this sewage contained nitrogenous matter enough to render fertile hundreds of acres of waste and barren land in the neighborhood of the city that was now lying idle and producing nothing. But the commissioners, who had spent several months in the examination of the subject, had not overlooked this subject. We find in the appendix of their report a description of the methods adopted to dispose of the sewage in fifteen modern European towns, in many of which costly experiments had been tried with the view to utilize sewage on the land.[18]

According to Philbrick, there was no doubt in the commissioners' minds that the only way to provide for efficient waste disposal was by carrying sewage with water to the ocean. "In short," he concludes, "the disposal of sewage by irrigation, though often the cheapest and best way of getting rid of it for cities remote from the sea or large rivers, could not be looked upon as a possible source of profit anywhere."[19]

Mansfield Merriman, in one of the first engineering sanitation textbooks to be published in the United States, *Elements of Sanitary Engineering,* fully endorses the water-carriage system and rejects any other methods as viable solutions for the American cities. "It was shown," Merriman writes in 1899,

> that the pail method for the removal of excremental matter is an offensive and impractical one, and, as it is not used in the United States, it will be classed among the public systems. The cesspool plan is also shown to be an objectionable one for a large town or city, and, although it is still extensively used in villages and country districts, it is to be regarded as a family method rather than a practicable and efficient public system and

hence will receive no further consideration. There remains, then, only the systems of removal by means of public water supply, known as the water-carriage systems, which are an outgrowth of the plan followed in ancient Rome. Had these ancient methods been continued and developed throughout Europe the thousand years of filth, disease, and misery known as the dark ages might perhaps have been a thousand years of cleanliness, health, and happiness.[20]

Innovative Unconventional Carrying System

A most interesting and often overlooked element of the debate was the actual design of the removal system. At the core of the technology was the use of piping for transporting waste. While advocates of the water-carriage system relied on ample water and water velocity to carry the waste, others pointed to alternative systems presently in use. These systems used vacuum, air, small pipes, and very little water to carry waste. These particular technologies allowed for better utilization of the organic matter and eliminated the need to expel harmful pollutants in existing water bodies. Advocates of these systems saw the potential for two separate methods for removing two types of waste: the removal of body waste by dryer conveyors and its composting in the earth, and the removal of gray-water waste from baths and kitchens utilizing existing stormwater drainage.

Such systems originated in Holland, where many towns were below sea level and therefore required sewer removal through pumping of some kind. The economical efficiency of these systems, together with the ability to make use of the organic matter as fertilizer, helped extend these systems to other countries such as Belgium and France.

American sanitary reformer George E. Waring Jr. offers one of the best accounts of this alternative system based on his visits to Amsterdam, Leyden, and Dordrecht in the 1870s. Utilizing small-diameter airtight pipes, and a mechanical vacuum system, each district of the city is connected by means of pipes and special holding chambers. Waring, describing the Liernur system, writes: "The initial principle of the system lies in the suction to a central public reservoir of the accumulation of famæl material deposited in receptacles at separate houses, these being connected with this reservoir by air-tight pipes.

The reservoir being exhausted of its air, the accumulations are drawn toward it by pneumatic pressure. No matter how large may be the area occupied by the sewered houses, each district has its central reservoir, and these reservoirs are in turn and in like manner themselves discharged into a main vacuum chamber at any convenient point, being connected with this by a similar system of pneumatic pipes."[21]

The vacuum system of Berlier, developed in France and used in some districts in Paris and Lyons, was based on similar ideas. Unlike the Liernur system, which applied vacuum suction at certain times of the day to transport waste to a central collection chamber, Berlier's system continually maintained partial suction in the street pipes, pulling waste from small individual collection basins placed in the cellar of each house.

In England, a similar technique—the Shone Hydro-Pneumatic system—was installed in a few towns, and like the Dutch innovation competed as an alternative to the water-based method.[22] Unlike the Liernur and the Berlier vacuum systems, the Shone technique relied on compressed air to push waste. Although it had been applied in a few cities in England, and even used to remove sewage from the House of Parliament in 1886, its most interesting debut was at the World's Columbian Exposition, held in Chicago in 1893. This was America's first international exhibition of the nation's vast technological and scientific strides made during the nineteenth century. It was the perfect showcase for the integration of innovative urban infrastructure with architectural beauty. (See figure 4.6.)

Located on a flat site near the shore of Lake Michigan, and dubbed "The White City," the fair represented an unprecedented collaboration of artists, architects, engineers, sculptors, painters, and landscape architects, who joined forces to create a single work—an ideal model city. Since Chicago had earned a national reputation as "typhoid-fever city" because typhoid, smallpox, and dysentery struck its population regularly, planners of the exposition site wanted to demonstrate an innovative system that did not pollute the lake with untreated water. Utilizing the Shone compressed-air technique, they removed sewage through clay pipes into ejector stations and precipitation tanks. These tanks, together with two water-treatment plants, were available for visitors to see as a sort of working exhibit that handled the sewer needs of the more than 6,500 lavatories and toilets, and the disposal of the sewage of 600,000 people per day.[23]

Figure 4.6

In England the Shone Hydro-Pneumatic system was installed in a few towns, and like the Dutch innovation, competed as an alternative to the water-based method. Unlike the Liernur and the Berlier vacuum systems, the Shone technique relied on compressed air to push waste. Although it had been applied in a few cities in England, and even used to remove sewage from the House of Parliament in 1886, its most interesting debut was at the World's Columbian Exposition, held in Chicago in 1893. (*Source: The Sanitary Engineer*)

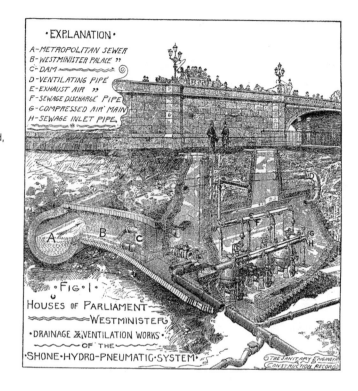

·EXPLANATION·
A—METROPOLITAN SEWER
B—WESTMINISTER PALACE »»
C—DAM
D—VENTILATING PIPE
E—EXHAUST AIR »»
F—SEWAGE DISCHARGE PIPE
G—COMPRESSED AIR MAIN
H—SEWAGE INLET PIPE

·FIG·I·
HOUSES OF PARLIAMENT
—WESTMINSTER—
·DRAINAGE & VENTILATION WORKS·
—OF THE—
·SHONE·HYDRO-PNEUMATIC·SYSTEM·

Despite favorable endorsements, calls for consideration, and the exhibit at the World's Columbian Exposition, these alternative systems failed to gain wider acceptance in the United States and England. Waring, for example, wrote that "all that it is safe to say about the system now, in its relation to our own condition, is that it is, as regarded in the light of what we know about the water system and the dry-earth system, sufficiently promising to justify the most energetic investigation. So far as I know, its opponents have adduced nothing against it that may not be remedied by practicable mechanical improvements, and its advocates, who are many, speak of its advantages with a confidence that, often at least, has grown from favorable experience of its practical working."[24] Yet most engineers and public officials viewed these alternative systems unfavorably. They objected to the need to build separate mechanisms for disposing human waste and kitchen (gray) waters. Unlike available drainage systems that relied on existing pipes, velocity, and gravity to do the work, these alternative vacuum and compressed-air systems were seen as complex and too technologically advanced for simple maintenance. Most importantly, sanitation engineers and public-health officials at that time believed in the purifying nature of water. For them, the quick removal of waste away from the city was vital. Any system that advocated the retention of waste and its utilization near human habitats was deemed dangerous and problematic. Such attitudes can be clearly seen in many of the reports issued by various sanitation committees of the time.

In 1880, for example, engineer Rudolph Hering, after returning from a sewer study tour in Europe at the request of the National Board of Health, wrote: "The general opinion held at present is that sewage must, beyond all other consideration, be disposed of in a way which is least injurious to the community, and that a pecuniary profit can not be looked for in every case. . . . When sewage can be safely discharged into a large river or the sea, this will generally be the most satisfactory and economical mode of disposing it."[25] Florence Nightingale, the British pioneer of nursing and the reformer of hospital sanitation methods, also saw water as the solution for proper sanitation. In one of her statements endorsing this new technology, she stated that "the true key to sanitary progress in cities is, water supply and sewerage. No city can be purified sufficiently by mere hand-labour in fetching and carrying. As civilisation has advanced, people have always enlisted natural forces or machinery to supplant hand-labour, as being much less costly and greatly more efficient."[26]

Ultimately, the water-carriage technology triumphed over the other systems. Since removal was the paramount concern and not the treatment or utilization of the waste, the authorities and the engineers involved concentrated solely on the transport system. Water carriage, they believed, would remove waste from habitation as fast as possible. It was simple, automatic, and did not require high maintenance. Such a system was attractive to the local authorities because it made waste disposal a more automatic procedure and relieved one from individual responsibility.[27] These views were widely expressed in various reports. "The lower classes of people," one said, "cannot be allowed to have anything to do with their own sanitary arrangements: everything must be managed for them."[28] (See figure 4.7.)

Employing a technology that could remove waste out of the sight and out of the mind of the public was the goal. As late as 1974 in an article published in *Scientific American* and titled "The Disposal of Waste in the Ocean,"

Figure 4.7
Water-based carriage systems require extensive infrastructure, such as this sewer construction in Brooklyn, New York. Historical decisions about the methods and systems for sewage collection have locked our current practice into a specific mode of operation. (*Source: Scientific American*)

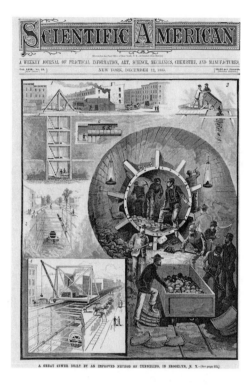

the author asserts that "contrary to some widely held views, the ocean is the best place to put certain wastes."[29]

Locked into a Technological Dependency

By means of a chain of past decisions we have established systems for managing our land, our towns and cities, and our water in ways that are proving to be an obstacle to progress.

Economists and political scientists often cite the dependency of present actions on past decisions in explaining certain equilibria in technological development. Brian Arthur, for example, applied a Polya process, a mathematical example introduced by Polya and Eggenberger in 1923, to demonstrate that random disturbances early in the history of selection, coupled with the self-reinforcing nature of these selections, shape subsequent outcomes.[30] Termed *path dependence,* this theory suggests that current equilibria, such as the state of technological development, industry location, or a firm's organization, is a function of early random shocks in the decision process that led to the equilibrium.[31]

S. J. Liebowitz and Stephen Margolis defined three distinct forms of processes that exhibit dependence on initial conditions. In the first form, initial actions create a path that cannot be left without a cost; however, the path proves to be optimal and the costs are being acknowledged. For example, a decision to use a particular type of aircraft may have a controlling influence for decades on an airline's fleet, but the long-term effects of the decision are fully understood and taken into account by the initial decision makers. In situations where the initial information is imperfect, decisions may not always appear to be efficient in retrospect. Under such second-form situations, the deficiency of the chosen path is initially unknown but is later recognized when the initial decision leads to regrettable outcomes and costly changes. In third-degree path dependence, initial conditions lead to an outcome that is inefficient, even though feasible options for modifications and improvements exist.[32]

Path-dependence theory has generated a lively debate in many circles. Examples include, among others, discussions about the qwerty keyboard layout, the use of VHS or Betamax video formats, and the utilization of Microsoft's Windows system in operating personal computers.

Arthur traces the decision regarding video formats and explains that dependency was created as a result of positive feedback in the video film-rental market. Video rental stores stocked more film titles for the system with a larger user base, and new providers chose the system for which they could rent more videos.[33]

Robin Cowan argues that political decisions have been the principal source of path dependence in the adoption of nuclear power techniques. He argues that the dominant "light-water" reactor design is inherently less efficient than potential alternatives, but this system was rushed into use because the Cold War political value of peaceful uses for nuclear technology overrode the value of finding the most cost-effective technique. Thereafter, engineering experience for the light-water technique continued to make it the rational choice for new reactors over less developed alternative designs, despite the fact that equal development of the alternatives might have made them superior. The principal U.S. suppliers and sponsors of light-water reactors, Westinghouse and General Electric, offered early systems at prices below cost in order to gain experience and offer improved systems to later users at higher prices.[34]

The archetypal case of path dependence has been, of course, the configuration of the typewriter keyboard. Paul David, for example, argues that the standard qwerty keyboard arrangement was dramatically inferior to other arrangements offered at the time. Since qwerty was the prevailing standard and most people learned that system, other layouts—such as the "Ideal" keyboard (1880) and the Dvorak keyboard (1932)—could never be introduced into the marketplace.[35]

Anthony Woodlief used path-dependence theory to show how city governments can get "locked into" suboptimal policies enacted at a time when the future consequences of these policies were unclear.[36] The history and choice of sewer technology is very much about path dependence and a "locked-in" situation. The adherence to a particular sewer-system paradigm in the latter half of the nineteenth century constrained innovative research and the implementation of alternative solutions. Sharon Beder puts forth a strong institutional argument that once a consensus is reached about a specific technology, and because this specific method is fully endorsed by a particular profession, it enables that group to consolidate their position and define themselves as the experts in that field. Other professionals are then likely to respect the boundaries of expertise set up by the particular technology, further entrenching it in practice. As standards are set and regulations established to support a particular practice, others are restricted to the role of outsider, or viewed as uninformed members of the public "in no position to question the range of treatment methods available."[37] Thus, the autonomy of the engineering community lay in its ability to dictate the range of technologies that would be taken seriously. And while "outsiders" could demand a different solution from within the paradigm, they were not able to ensure that alternatives from outside the paradigm would seriously be considered.

Water-carriage systems, as advocated by sanitary reformers and government authorities, required an integrated system of underground pipes that was to be planned, engineered, and coordinated with reference to a larger, citywide plan. Political boundaries could not fragment a sewerage scheme; rather, local councils were forced to give authority to more centralized governmental bodies in the realm of waste disposal once water-carriage systems were adopted.

Water carriage, with its economies of scale, created the need for central administration, and thus was an important factor in facilitating government integration. Since water-carriage technology involved large-scale construction and required the centralization of methods of operation, it also brought sewer disposal within the engineering domain and was favored and quickly endorsed by many engineers.

The reform measures pushed by sanitary reformers in the nineteenth century were largely technological, and the development of new technologies associated with water supply and the water carriage of sewage offered the opportunity for a new professional group to form that claimed to have specialized knowledge in the field. Engineers were closely associated with large-scale public works, including the construction of tunnels and the laying of pipes, and overseas engineers were carving out a profession for themselves in the area of sanitation.

Water-carriage systems and the use of centralized wastewater-treatment plants had great appeal for policymakers and those in the engineering profession.[38] Because the laying of sewer pipes and the building of sewage-treatment plants are costly endeavors, and since typical consultants' fees can be as high as 20 percent of the estimated construction cost, there was little incentive in finding and endorsing low-cost solutions.

Modifying Trends

In 1972, amendments to the Federal Water Pollution Control Act were passed by the U.S. Congress. Section 201 of that act provides subsidies for the construction of local treatment works, both treatment plants and the interceptor sewers that serve them. At that time, $18 billion was authorized by Congress, providing 75 percent funding to both new and old communities for approved projects.

This act created one of the largest capital facilities programs in the nation's history. Its effect on urban land use and the institutionalization of a specific technological standard has been as dramatic as the changes brought by the Federal Highway Program.[39] Furthermore, research shows that in the initial physical design phase of sewer systems, regulations and engineering procedures seem to encourage, or at least permit, the design of unnecessarily large and extensive sewerage systems. In numerous projects, design engineers predicted tremendous increases in the service area's population, sized the sewer lines with a great deal of excess capacity to accommodate this growth, and designed systems large enough to serve the maximum ultimate density of population projected for the area.[40]

The standard per capita wastewater-flow formula employed in sizing sewer interceptors is also creating unnecessary capacity. Commonly, the Environmental Protection Agency and other governing agencies encourage engineers to design sewer lines and sewage-treatment facilities based on a standard wastewater-flow figure of 100 gallons per capita per day (gpcd). However, as a recent survey shows, in some regions of the country, this standard was recently raised to 125 gpcd for tentative sizing of a treatment plant's capacity. This is even more troubling, because most engineers surveyed agreed that the 100-gpcd figure is generous and was higher than actual per capita water consumption. In fact, in some rural areas, the actual consumption has been documented at half that figure.[41]

Adding to the problem, many environmentalists and public-health officials, in the 1970s, endorsed centralized sewage-treatment systems, because they saw in them a cure for waterway pollution. With the Clean Water Act of 1977 and the establishment of standards that support water-carriage-based systems, all practices were locked into a particular mode of operation by law.

Outside the mainstream, innovators are working on various ecological methods for the utilization and recycling of sewage. In 1996, for example, a system called the Living Machine was installed in South Burlington, Vermont, in order to determine the effectiveness of ecologically engineered wastewater systems. These independent and self-contained biological treatment systems consist of constructed wetlands, which filter and clean wastewater. They require little piping, and their on-site location eliminates the need for an expensive carrying system and treatment facility.[42] (See figure 4.8.)

Under the auspices and support of the Environmental Protection Agency, 80,000 gallons of municipal sewage were diverted and treated by the Living Machine. Since the main purpose of this project was only to test the process, the EPA and the city of South Burlington would not allow the treated waste from the system to be directly discharged into the waterways. Even though an analysis performed by the EPA found that the end product of the Living Ma-

Figure 4.8
Alternative systems such as this Living Machine do not conform to existing standards and therefore cannot easily be implemented in public projects. (*Source:* Courtesy © Dharma Living Systems)

chine met or exceeded the water-quality standards of traditional systems, both the EPA and the city insisted that the water be cleaned again in the traditional treatment plant.[43]

Although the Living Machine experiment showed that it could match traditional wastewater systems, the initiative has not materialized beyond the experimentation phase. Once funding ended, neither the city nor the neighboring industries showed any interest in continuing its use. South Burlington stated that they could not afford to operate what were essentially two wastewater-treatment plants; the Living Machine was not capable of handling large quantities of city's waste, and therefore the continued operation of the traditional treatment plant was still required.

Alternative on-site wastewater-treatment systems such as the Living Machine have great potential for creating a paradigm shift in the treatment and transport of wastewater. Even the EPA has realized that these systems can act as more than temporary solutions and can serve as low-cost, environmentally friendly alternatives to traditional systems. In a 1997 report to Congress, the EPA stated that "adequately managed decentralized wastewater systems are a cost-effective and long-term option for meeting public health and water quality goals, particularly in less densely populated areas."[44] Yet most local regulations limit the introduction and use of such alternatives. It is difficult for prescriptive codes to specify the full range of technological options appropriate for a given site and anticipate the different sensitivities of the site's water and land resources. Most regulatory barriers to on-site wastewater-treatment systems are derived from the perception that the decentralized nature of these systems prevents the necessary oversight to ensure good water quality. Tight regulations make it difficult for developers to implement alternatives. Municipalities also find that funding opportunities play a large role in their decision on whether or not to implement alternative systems. Many funding and planning grants are focused on the traditional systems, making them more economically beneficial for municipalities to build.

While government initiatives in the area of regulatory reform may be slow to occur, industry forces such as developers, mortgage companies, and homeowners' insurance programs may create positive change. An interesting example has germinated in Washington State through an evaluation of accountability. This effort involves an attempt to upgrade the alternative on-site sewer industry through an insurance program that gives every practitioner a stake in the success of each system by assuring the owner that the alternative system will perform as designed.

Such accountability by the industry allows the designer, the manufacturer, the installer, the operator, and above all the public regulator to operate without the fear of liability. In fact, the National Onsite Wastewater Recycling Association (NOWRA), which advocates this approach, is ready to negotiate the accountability agreements with insurers to monitor success, as well as to negotiate coverage rates for practitioners based on performance.[45]

By letting the industry take responsibility through insurers, the onus on the regulator would shift from enforcement to monitoring to code modifications. With sets of local operation data, and a shift from national to regional testing performance, performance standards suitable to each locale would evolve. These standards would be designed to address specific site and watershed goals.

The development of such performance standards, together with innovative accountability approaches, would provide the tools for communities that are willing to build alternative systems. Ultimately, such approaches may need the backing of various government agencies, because local officials are often reluctant to approve the use of new technologies locally when no endorsement is given by higher authorities.[46]

Changes in sewer-system designs seem to come about only when circumstances force alternatives to standard procedures. Christine Rosen, in her work on the fires that devastated Baltimore and Boston, shows that infrastructure-system designs were the last to adapt to imperatives of economic forces and social reforms, and their modernization was typically related to their failure or underperformance.[47]

David Wojick, in the "Structure of Technological Revolutions," argues that anomalies occur in technological paradigms when standard procedures repeatedly fail to eliminate known ills or when knowledge shows up the importance of factors that have previously been incorrectly evaluated.[48] Yet in many instances, those contesting existing practices may be outside the paradigm community, and their views are often disputed. The government regulatory authorities are unlikely to force changes because they are well aware of the costs involved in changing the system. For these new developments to be incorporated into standard planning practice, a change in the engineering and urban planning paradigm needs to be made, in particular in the current emphasis on "good enough" solutions based on slide-rule answers from often obsolete formulas. Both public officials and engineers must realize that the science of sewer design is capable of change. The fruit of alternative research must be accepted and converted into practice supported by flexible guidelines.

5

Regulating Developers

While regulations are intended to guard against the evil results of igno-
rance and greed on the part of landowners and builders, they also limit
and control the operations of those who are neither ignorant nor greedy;
and it is clear that the purpose in framing and enforcing them should be
to leave open the maximum scope for individual enterprise, initiative and
ingenuity that is compatible with adequate protection of the public interests. Such regulations are, and always should be, in a state of flux and
adjustment—on the one hand with a view to preventing newly discov-
ered abuses, and on the other hand with a view to opening a wider opportunity of individual discretion at points where the law is found to be
unwisely restrictive.

—Fredrick Law Olmsted Jr., 1916

To the private-sector, professional consultants as well as some public officials,
Fredrick Law Olmsted Jr.'s statement made almost a century ago still holds
much truth.[1] Many are still apprehensive about the effect of development-
related regulations on their practice. They often see regulations as being costly,
inconsistent, and superfluous, and perceive codes and standards as a barrier to
housing affordability and innovative design solutions.

In *Ecological Design*, Sim Van der Ryn and Stuart Cowan write that "city
planners, engineers, and other design professionals have become trapped in
standardized solutions that require enormous expenditures of energy and
resources to implement. These standard templates, available as off the shelf
recipes, are unconsciously adopted and replicated on a vast scale. The result
might be called dumb design: Design that fails to consider the health of human
communities or of ecosystems."[2]

Like Van der Ryn and Cowan, others have also called for regulatory reforms and alternative solutions to bring better design resulting in efficiency and site suitability. Albert Bemis, writing in 1934, asserted that "compliance with minimum standards with respect to street grading and the installation of water mains and sanitary sewers often may increase the total home cost as much as 20 percent."[3] Jesse Clyde Nichols, who in 1906 established the famous Country Club District in Kansas City, declared that "the building codes of many of our cities are obsolete, drawn to favor certain industrial trades and certain types of merchandise which create unnecessary cost of home construction."[4]

The modern process of regulation of human settlements began with the nineteenth-century urban public-health crisis, when decisions were made to create improved public water and water-carriage sewer systems. Related issues such as jerry-built structures, patchwork subdivisions of tangled property lines, and broken street alignments resulted in parallel movements for building codes, street surveying, and, ultimately, twentieth-century use and structure zoning and subdivision controls. By 1950 the land-development rulebook embraced every aspect of the physical design of neighborhoods, and through its detailed requirements it profoundly affected many significant social and economic issues as well.

Amidst a housing boom, a 1964 survey found that builders cited finance, labor, merchandizing, and material costs as the major obstacles to new construction. A dozen years later government regulations and the lack of suitable land pushed all of these earlier issues into the background. (See figure 5.1.)

This dramatic shift from perennial builder complaints about the enduring cost elements of the industry to government laws and administration had its roots in the awkward grafting of new public ambitions on old regulatory frameworks and methods. Since the 1964 survey of builders, the momentum of the civil rights movement added a new set of concerns about how municipalities strengthened or weakened racial and class segregation. Today's "affordable-housing" issues are legacies of those earlier initiatives. The environmental movement of the 1970s and subsequent years added a fresh set of demands. Federal mandates, such as the Disabilities Act, the Clean Water Act, and energy conservation standards, added further directives. Many of the new subjects for regulation lengthened the list of boards and authorities that must

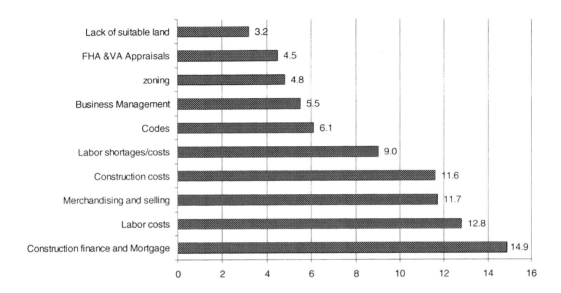

Lack of suitable land	3.2
FHA &VA Appraisals	4.5
zoning	4.8
Business Management	5.5
Codes	6.1
Labor shortages/costs	9.0
Construction costs	11.6
Merchandising and selling	11.7
Labor costs	12.8
Construction finance and Mortgage	14.9

be consulted by anyone contemplating a new development. Finally, metropolitan municipalities added still more demands to the rulebooks as they tried to ease their fiscal burdens by shifting public infrastructure costs onto new private developments.

Figure 5.1
National Association of Home Builders Survey of Significant Problems in 1964, percent distribution. (*Source:* Seidel 1978)

A Balancing Act

Since many of the new demands depended on the balancing of physical, social, and natural systems, they did not lend themselves well to the traditional engineering specification methods. Gallons per hour, floor-area ratios, building heights, and simple use categories of residence, industry, and commerce were ill-suited to the evaluation and design process decisions now required. Specific place-based judgments would have been more appropriate to the new regulatory ambitions rather than a reliance on universal codes.

One way to better understand the significance of the changes in the American regulatory climate over the past half century is to follow the reports of the two principal actors: the regulators and the regulated. In 1952 the regulators were on the defensive, seeking to enlarge their role against the pressure of private developers. By 1976 the positions had reversed themselves. The developers now felt harassed, while the regulators flirted with ideas of reforms.

In its 1952 manual, the U.S. Housing and Home Finance Agency pressed for more widespread subdivision controls: "The regulation of land subdivision for residential and other uses is widely accepted as a function of municipal and county government in the United States. It has become widely recognized as a method of insuring sound community growth and the safeguarding of the interests of the homeowner, the subdivider, and the local government."[5] Two years later the American Society of Planning Officials warned planners about the home builders' "campaign to break municipal subdivision regulations and controls" and their intent to pressure municipalities "to abandon or weaken subdivision control ordinances, financial regulations and control."[6]

The planners needn't have worried. At the same time, the influence of the federal government's mortgage lending guidelines, and the need to ensure public investment, brought on a wave of municipal and state regulations. This proliferation of regulations soon called forth a flood of studies examining the effects on design, housing costs, and the socioeconomic patterns of neighborhoods. In the end, the studies concluded that the accumulation of rules and regulations had become dysfunctional.

As recently as 2003, a study by the Pioneer Institute for Public Policy Research and the Rappaport Institute for Greater Boston concluded that in Massachusetts, "Excessive regulation by agencies and boards at both the state and local level has gotten to the point of frustrating the development of housing. . . . Both levels of government need to prune back the sprawling regulations and improve coordination among the different regulatory players."[7]

Two studies from 1976 and 2002 report on the new, reversed position of regulators and regulatees. In both these years government regulations and their modes of administration loomed as the most significant barriers to the pathways appropriate to development.[8] (See figure 5.2.)

The phrase "imposed regulations" summarizes a cluster of complaints by developers. In this category they focused on the lack of coordination among agencies as well as unnecessary delays. Also, since 1976, builders have complained of "unnecessary costs," and there is evidence of class-exclusionary practices among well-to-do municipalities. Developers frequently offered comments such as the following:

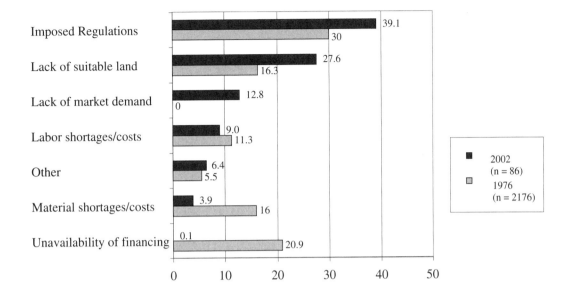

Imposed Regulations — 39.1 (2002), 30 (1976)
Lack of suitable land — 27.6 (2002), 16.3 (1976)
Lack of market demand — 12.8 (2002), 0 (1976)
Labor shortages/costs — 9.0 (2002), 11.3 (1976)
Other — 6.4 (2002), 5.5 (1976)
Material shortages/costs — 3.9 (2002), 16 (1976)
Unavailability of financing — 0.1 (2002), 20.9 (1976)

■ 2002 (n = 86)
□ 1976 (n = 2176)

"Regulatory agencies exceed their authority to practice social engineering, architecture, and micro-management."

"Subdivision codes don't allow any flexibility. They are too standardized. More flexibility in subdivision codes is desperately needed."

"City and county offices have no sense of fairness. They are only interested in exactions and imposing regulations that make them appear more successful in protecting the community from the 'evil' developer that may be trying to be profitable."[9]

Of course, not all regulations are perceived as equal, or even detrimental to development. In trying to understand the relationship between different regulations and their impact on development, the two surveys asked respondents to indicate the type of regulations that increased the final selling price of a unit by 5 percent or more. Subdivision regulations and building codes clearly stand out as the dominant force that impacts new developments.[10]

The justification for governmental imposition of subdivision controls is rooted in the police power—the right of political entities to regulate in order to promote the health, safety, and general welfare of the community. As such, three general goals can be seen in the establishment of such regulations:

Figure 5.2
Housing industry, significant problems comparison, in 1976 and 2002 (weighted-scale selection). (*Sources:* Seidel 1978, Ben-Joseph, 2003).
Note: 1976 data is based on a 3, 2, 1 weighted scale with totals divided by a factor of 6; 2002 is an average of respondents three, nonscaled, selections.

- Preventing premature partial subdivisions that are poorly linked to the broader community
- Preventing poor-quality, substandard subdivisions with inadequate public facilities and infrastructure
- Reducing financial uncertainty and risk to the investor, buyer, and community

In his 1976 study Seidel points to two important factors resulting from these practical goals: the exclusionary implications of subdivision regulations, and the hidden increase of cost due to a prolonged approval process. With regard to the exclusionary aspect, Seidel writes: "The desire to ensure high-quality subdivisions is sometimes synonymous, in effect if not always in intent, with the exclusion of those people who can afford only low-cost housing. Thus any rationale for extensive subdivision requirements justified on the basis of avoiding 'blight' demands more than superficial inspection. The level of public improvements required must be scrutinized to determine whether or not the regulations are actually designed to erect an economic barrier to keep out the poor and, increasingly, those with a moderate income as well."[11]

Prolonged administrative and approval processes required in the administration of subdivision regulations not only increase the financial risk for the investor/developer but also increase the cost to the home buyer. According to Seidel, for every additional month added to the completion date, there is a 1–2 percent increase in the final selling price of the unit.[12] With the recent survey indicating a steady increase over the last 25 years in the average time it takes to receive subdivision approval, the increase in cost has undoubtedly been transferred to the consumer.

The prolonging of approval times can be understood by measuring the new complex packages of regulations as they affect the specifics of approvals. A convenient way to locate the sticking points is to examine in turn the administrative process, the site and design requirements, and the relationship between subdivision controls and other regulations like growth controls.

Subdivision-Approval Process

Procedures for subdivision approval have followed the standards established by the FHA in the late 1930s and early 1940s.[13] These are based on three main stages: preapplication, conditional approval of preliminary plat, and final plat approval. In the preapplication stage, the subdivider gathers the information and data on existing conditions, studies the site suitability, and with the help of professionals, develops a preliminary plan in sketch form to be submitted to the planning commission for advice and assistance. The planning commission reviews the plan in relation to a master plan, design standards, and improvement requirements, and notifies the subdivider of their issues and concerns if any.

In the second stage, the subdivider, if opting to develop, submits a revised preliminary plat for conditional approval by the planning commission. Once the plan is approved, the subdivider stakes out the plat according to the approved preliminary plan, and either installs improvements or posts bonds to guarantee completion of improvements. A final plat is then submitted for final approval. Once the planning commission approves the final plat, the new plats are recorded and development begins.

While the original FHA guidelines seem simple and straightforward, the realities of the last decades are those of growing complexity and frustration of those involved in the process. Indicative of these frustrations is the following statement by the Urban Land Institute: "American developers of housing must deal with an expanding array of regulations at every level of government. Unreasonable regulations on development inevitably inflate paperwork required for a project and intensify the complexity of data, analysis, and review procedures for both public and private sector. Ultimately, the delay caused by the regulatory maze produces higher cost housing through holding costs, increased expenses due to risk, uncertainty, overhead, and inflated cost of labor and materials, and other more hidden costs."[14]

Some of the blame for the costs of the approval process should be placed on the rigidity of its steps. The progress from sketch plan to preliminary plat approval, to terms and conditions approval, to final approval does not allow for easy and quick revisions. Perhaps more significant delays arise from the increasing numbers of agencies and committees that must pass on the developer's proposals.

Almost all public officials surveyed (97 percent) laid the blame for approval delays on the developers. In their judgment developers do not provide sufficient information about proposed developments, and are often changing plans. Such an assessment clearly indicates that a lack of good coordination and communication between developers and public officials is a major problem.

Nevertheless, some of the blame also can be attributed to the approval process itself. More than half of the public officials surveyed recognized that delays were caused by inefficient management and lengthy approval processes by other agencies and commissions. They indicated that in more than 40 percent of the cases at least ten other agencies (beside the planning commission) took part in the approval process. Topping the list were municipal sewage and health departments.

Time and Delays in the Approval Process

Delays and a prolonged approval process not only are prohibitive to a developer, but also have cost consequences for the consumer. In most jurisdictions surveyed (42 percent), the average time period between initial submission of a (typical) subdivision application and tentative (or preliminary) approval is 2 to 4 months. In 34 percent of the cases, approval takes less than 2 months. Although these numbers indicate an efficient turnaround, it should be noted that overall there is some decline in efficiency as compared to the 1976 survey. For example, in 1976 half of the jurisdictions surveyed approved a preliminary plat in less than 2 months, 47 percent approved rezoning in less than 2 months, and 33 percent approved variances or special relief in less than 1 month. In 2002 only 27 percent of the jurisdictions surveyed were able to grant rezoning in less than 2 months and only 14 percent allow for variances.

Unlike the public officials, developers reported very different estimates of the time it takes to obtain approvals. According to the developers surveyed, it took on average 17 months in 2002 to obtain all the required permits. This lengthy approval time is consistent with the findings from Seidel in 1976. In both 1976 and 2002 the majority of the developers surveyed, 47 and 45 percent respectively, received all approvals for development between 13 and 24 months. The percentage of developers indicating that they received all approvals in less than 7 months declined in 2002 by almost half in comparison to 1976. Furthermore the number of those reporting it took over 2 years to get approvals doubled in 2002 to 20.5 percent.

Discrepancies can also be seen in the estimated time required for granting variances and zoning relief. According to the majority of the developers surveyed, it took more than 4 months to obtain variances, special exceptions, or rezoning. The majority of public officials, on the other hand, indicated an average of 1 to 2 months for variances, and 3 to 4 months for rezoning.

The discrepancy in time estimates between public officials and developers may be explained by their respective views of the development process. While public officials see timely approval as a yardstick for measuring public performance and service, developers see each delay as part of an unnecessary bureaucratic process. Another possible explanation has to do with the frequency and length of time by which special variances and zoning relief are being processed and approved. As noted previously, most public officials indicated that when such measures have to be taken, approval of the relief itself can take on average between 3 and 4 months.

Interestingly, the time it takes to get an approval is much shorter in low- and moderate-income communities. Above 80 percent of these jurisdictions approve subdivisions in less than 5 months, as compared to 60 percent of the higher-income jurisdictions. Although a lengthier approval process in middle- and higher-median-income communities may indicate a more detailed and comprehensive approval process, it can also indicate that delays and length may be used as a tactic to exclude development.

Excessive Design Standards

Excessive street and right-of-way widths, rigid earthwork specifications, and overdesigned infrastructure systems are unfavorable to the introduction of site-sensitive solutions, and often impede cost reductions. For example, the right-of-way width for a residential-subdivision street, as specified by the Institute of Transportation Engineers, has remained at 50 to 60 feet for at least 40 years.[15] Such ample space, designated for an exclusive monofunctional land use within a residential environment, has contributed to the perception that in its present form, the typical subdivision is grossly wasteful in its use of energy, material, and land.

A 2002 study by the American Rivers, the Natural Resources Defense Council, and Smart Growth America shows that wide streets, excessive parking requirements, and increased pavement around setbacks contribute to loss of potential groundwater infiltration.[16] Subdivision sewer-system standards are also so entrenched and widely accepted that alternative planning, sizing, and location of the systems is seldom considered.

Developers clearly expressed their frustration with the excessive and often unwarranted nature of the physical improvements and standards associated with subdivision development. When asked to indicate which requirements present the greatest expense, in conforming to regulations, an overwhelming majority (80 percent) pointed to requirements associated with site design.

When asked to indicate which requirements they perceived as excessive, 52 percent of the respondents indicated requirements relating to street construction, with 45 percent indicating land dedication, and 43 percent storm-sewer piping (underground piping for stormwater mitigation). When asked to indicate more specifically which physical standards within each category were seen as excessive, the most frequently cited were: street widths (75 percent of the respondents), street rights-of-way (73 percent), and requirements of land for open space (73 percent). (See table 5.1.)

While one might expect that developers would criticize regulations and see them as interfering with their business, it is important to note that most respondents were selective in their answers to the survey. Out of twenty-nine requirements listed, only thirteen were seen by the majority of the developers as excessive, while sixteen others seemed reasonable. Such a distribution indicates that many developers are in tune with construction and design performance and that their attitudes toward regulation cannot always be assumed to be negatively biased.

Furthermore, our surveyed public officials (town planners and town engineers) have often concurred with the developers' observations. Generally, these officials agreed that aspects of the regulatory process, such as the enforcement of subdivision regulations, have become more demanding and complex. For example, over the past 5 years, 70 percent of the jurisdictions where these public officials work have introduced new requirements, and 57 percent have increased specifications, such as those for setbacks and lot sizes. Only 16 percent of these jurisdictions have decreased their specifications, mostly by reducing street widths.

Seeking Relief

Government regulations, particularly those pertaining to the design and control of subdivisions, are seen by two-thirds of residential developers as the main culprit in prohibiting design innovation and increasing the cost of housing.

Table 5.1 Developers' assessment of various requirements (percent respondents, n = 79)

Requirement	Excessive	Not excessive	Standard error of estimate
Street width	75		4.5
Street ROW	73		4.6
Pavement thickness		62	5.2
Curbs		83	4
Sidewalk width	56		5
Sidewalk thickness		70	4.7
Water-pipe diameter		55	5
Water-pipe material		80	4
Water-pipe depth		93	2.6
Water-pipe hook-up fees	85		3.7
Sewer-pipe diameter		72	4.6
Sewer-pipe material		75	4.5
Sewer-pipe depth		70	4.7
Sewer hook-up fees	90		3
Sewer-system layout		56	5
Stormwater-pipe diameter	62		5
Stormwater-pipe material		50	4
Stormwater-pipe depth		45	5
Stormwater-pipe hook-up	57		5
Stormwater-system layout	73		4.6
Street trees	73		4.6
Street lighting		52	4
Telephone lines		53	4
Electric lines	60		5
Cable/TV lines		64	5.4
Land for recreation	52		4
Land for open space	73		4.6
Land for schools		65	5.4
Fee in lieu of land	79		

More specifically, they see these regulations as an impediment to efforts to increase densities, change housing types, and reconfigure streets and lots.

One way developers try to relax these regulations is through zoning relief and variance requests. Indeed, more than half (52 percent) of the developers surveyed had to apply for some sort of relief in at least half of their projects, while 37 percent had to apply in at least three-quarters of their projects. When asked to point to the type of changes they applied for, many indicated they applied for variances that would allow them to build higher-density single-family projects, and include more multifamily units. They also would create more varied site and structural plans if they had the opportunity. Seventy-two percent indicated that because of existing regulations they had to eventually design lower-density developments than they had intended.

Similar findings by Levine and Inam show that 78 percent of developers nationwide view local regulations, including zoning, subdivision regulations, and parking standards or street width, as a significant obstacle to the creation of developments with higher densities, mixed use, and transit-oriented design. According to Levine, although developers perceive considerable market interest in such forms of development, and believe there is an inadequate supply of such communities, they also believe local regulation is the primary obstacle to their construction.[17]

These findings should alarm individuals dealing with housing reform, as well as those who, as early as the 1970s, warned of the consequences of various exclusionary devices. Restrictions introduced in the 1950s against higher-density developments, multiple housing types, and minimum lot sizes and floor areas are still hampering the housing industry. Developers in both 1976 and 2002 felt subdivision standards and zoning regulations increased the cost of the homes they built and decreased densities. In many instances these regulations pushed developers to build in green-field locations, away from major urban areas, where restrictions and abutters' objections would be less severe. (See table 5.2.)

Toward Better Subdivision Regulations

The mazes of codes, regulations, and design requirements placed on residential developments have often been at the center of contention between developers and public officials. At the core of this friction may be the simple fact that

Table 5.2 Effect of local regulations on developers

Effect of local regulations on developers' decisions	Percent respondents, $n = 85$	Typical relief sought by developers in the majority of their applications	Percent respondents, $n = 83$
Build less dense development than originally desired	72.1	Denser single-family housing	42.4
Build more expensive structure than originally desired	60.6	Lot-size decrease	39.7
Build in less populated areas	38.5	Include or change to multifamily housing	31.7

many subdivision requirements imposed today have little to do with the rationale that shaped them in the early twentieth century. Health and safety concerns caused by inadequate building and infrastructure construction, premature subdivision of the land resulting in conflicting property lines and neighborhood layouts, and builders who were not concerned about their reputation have little bearing on present realities.

Regardless of the numerous calls for regulatory reform, changes to subdivision controls have been slow. Indeed, as Seidel's and our study indicate, over the last 25 years the subdivision-approval process has increased in its complexity, in the number of agencies involved, in the number of delays, and in the addition of new requirements.

In the instances where our study examined the universe of assorted regulations according to the median income of the communities surveyed, the results show that higher-income communities provide fewer options for performance guarantees, require higher dedication of open space from the developer, and generally are the ones to implement growth-control measures. Although the sample is relatively small, such indications suggest that exclusionary tactics in these higher-income communities may be more prevalent than is often assumed. Interestingly, a study has shown that two progressive Massachusetts laws, both of which should give developers and communities tools to build affordable housing—Chapter 40B, known as the Comprehensive Permit Law or "antisnob zoning" law, and the Community Preservation Act—have actually become instruments for antihousing sentiments and actions.[18]

Under such conditions, change is unlikely to happen through traditional means but rather is likely to be pushed through by nonconformists. Indeed, in the last decade most innovation in subdivision design has evolved within the private domain and under the governance of community associations. Two such innovations, new urbanism and conservation (or green) subdivisions, would not have been possible if it were not for early prototypes such as Seaside, Florida, and Prairie Crossing, Illinois—communities built as common-interest communities (CIC), privately owned and maintained by homeowners associations.

Renegades such as these CICs often serve as catalysts in changing subdivision standards and regulations. To diffuse such innovations in subdivision design and planning, public officials together with agents of the housing industry must move beyond confrontation into joint association. It is essential to continue studying and documenting the impact of engineering standards and codes, such as those relating to street widths, rights-of-way, and building setbacks, on residential-development forms and housing costs. Public officials should evaluate federal land-use policies, such as those associated with environmental regulations, that hinder design changes to subdivision patterns, form, and density.

The red tape and bureaucratic procedures associated with development approval at the local level also result from the involvement of multiple agencies and committees in the process. To eliminate delays and jurisdictional conflicts, localities can consider consolidating this process into the hands of one agency, and establishing a uniform structure for appeals to be reviewed and approved by this sole agency. Streamlining the process can also be improved by introducing electronic permitting systems. As Internet use is spreading, there is a growing expectation of being able to conduct affairs from home or office with greater immediacy. From automatic approval of plans, to equipping inspectors with portable devices for recording and inspecting, electronic permitting systems can provide better and more timely information to decision makers and experts alike. The possibility of electronic plan review is particularly encouraging for its potential to automatically analyze a plan, and compare it with codes and standard requirements. Alternatively, such systems can allow the plan reviewer to enter various descriptors and benchmarks, and let the software call up the applicable requirements that need to be considered. The process can ease the burden of subdivision planning and ensure a certain consistency of performance for many towns with limited or no planning staff. A recent HUD report that strongly supports such systems in the effort to reduce regulatory barriers to housing, indicates that in jurisdictions that have implemented such systems, turnaround time was reduced by as much as 80 percent.[19]

Obviously there are many issues to tackle in shaping a new regulatory template for subdivisions. But none is more important than the realization that this new template must allow and promote a variety of housing styles and types of development design. In the last few decades decisions regarding the built environment were often made by those far removed from understanding design and its impacts. The planning profession has generally been reluctant to champion physical design, largely because of an ideological commitment to social science–based disciplines as the foundation for urban planning education and practice. This has resulted in the marginalization of urban design and physical planning to the point that it all but disappeared from urban planning curricula. Physical planning tasks have been turned over to others following the formulas of local codes and regulations. This has not only created a one-dimensional approach to planning, but it has also rendered planning practices inadequately prepared to deal with current environmental and development trends.

The increased prominence of ecology, sustainability, and associated lifestyles has brought physical planning and design to the fore. The question of how subdivisions should be planned to minimize their ecological footprint and impact, has gained renewed importance. A renewed emphasis on place and ways of living has brought urban design to the forefront.

All in all, the restored emphasis on physical planning has exposed the inadequacies of common regulatory mechanisms. This renewed bond between design and planning, between shaping space and its context, between the expert and the community, presents new opportunities. Planners, architects, and engineers can now challenge existing regulatory practices based on the poor performance of these practices, they can provide place-based criteria responsive to the local and not the universal, they can streamline an exhaustive process, and they can turn obscurity into a clear vision that communities can grasp.

6

Second Nature

Every owner or inhabitant of any and every house in Philadelphia, New-castle and Chester shall plant one or more trees, viz., pines, un-bearing mulberries, water poplars, lime or other shady and wholesome before the door of his, her or their house and houses, not exceeding eight feet from the front of the house, and preserve the same to the end that the said town may be well shaded from the violence of the sun in the heat of summer and thereby be rendered more healthy.

—Pennsylvania Shade Tree Law, 1700

In the spring of 2002, contractors were busy at work in a rain-soaked, 40-acre Louisiana rice field. The site for the proposed Jennings High School complex, the once waterlogged rice field, would include about 255,000 square feet of buildings, a gymnasium, parking, and sports fields. Built 2.5 feet above the surrounding cropland, the construction needed a little ingenuity and a lot of soil replacement. "We essentially took off about 2.5 feet and came back with 5 feet of new soil," Sam Cavys of the construction corporation said proudly.[1] The appropriateness of building the school on this flood-prone rural site was never an issue. Rather, the driving force was the availability of the undeveloped land and the ability to mitigate its natural constraints through engineering.

Removing and filling earth, or draining wetlands with pipes and channels, may allow for our built environment to expand almost indefinitely. But while this land conversion may be unstoppable, its ensuing physical pattern, derived from the application of construction techniques, design manuals, and planning regulations, must be questioned.

Could one simultaneously protect and enhance natural features, reduce development and infrastructure costs, and increase the marketability of a built project? The answer is probably yes if one chooses to integrate into the development process alternative planning solutions fitted to the site's unique features. These may include unconventional stormwater-management techniques, minimal grading and earthworks, and the substitution of centralized infrastructure systems with on-site options. Yet the road to implementation is full of challenges. From restrictive local ordinances, to rules and regulations that no longer reflect today's knowledge, to ingrained construction practices, change is slow to occur.

Water and Site Integrity

One of the drastic changes caused by development is its effect on soils, vegetation, and water. Obviously, with the increase in impervious surfaces, stormwater volume and speed increase. A one-acre parking lot, which produces almost sixteen times the volume of runoff that comes from a meadow of the same size, not only increases water volume but also prevents recharge of the aquifer. When increased volumes of surface water reach existing streams, the shape of the stream channel deteriorates rapidly. Streams in such urbanized areas often lose their ragged edges, small natural pools, and woody debris, resulting in spawning decline. As runoff flows across paved roads and parking lots, water temperature rises and pollutants such as oil, metals, and soils are carried into streams and waterways. Consequently, a decrease in oxygen levels and an increase in nitrogen reduces the thresholds needed to keep a habitat healthy.

Excessive impervious surfaces and piped drainage systems also pose a danger to our supply of potable water. A joint study by the American Rivers, the Natural Resources Defense Council, and Smart Growth America showed that in some of the largest metropolitan areas, the potential amount of water not infiltrated into the ground annually ranged from 14.4 billion gallons in Dallas to 132.8 billion gallons in Atlanta. Atlanta's "losses" in 1997, for example, amounted to enough water to supply the average daily household needs of 1.5 to 3.6 million people per year.[2]

Impervious surfaces continue to increase due to land-consumptive requirements such as large extensive road frontages and wide, paved roads. Many of these streets are not only paved with watertight concrete and asphalt, but they also support removal infrastructure. This engineered system is built to remove water as fast as possible from any urban site to the nearest stream by curbs, gutters, and pipes. As seen in chapter 4, this type of removal infrastructure underwent a rapid change with the sewering of cities through the use of pipes. With much of the historical debate focusing on fast and efficient removal, a consensus was reached at the end of the nineteenth century that piping should be the ultimate conveyance method. With slide rules and formulas for calculating storm intensity-duration-frequency and water velocity, pipes, concrete channels, curbs, and gutters become the eventual choice for removing all types of liquids—including rainwater.

Yet the consequences of efficiency and speed were downstream flooding, massive erosion, and loss of valuable property and land. Calls for amending drainage practices had their influence in the 1970s with the establishment of ordinances for on-site water detention. The idea was simple: hold the extra surface water generated by development, and release it after the storm was over. This was best implemented through the design of detention ponds.

Systems of ponds have sprouted all over the country. Hoping to mitigate the ever-increasing developments with their impervious surfaces, localities demanded through rules and regulations the construction of detention systems. Yet as more of these ponds were constructed the more flooding continued to happen. Except this time, the floods lasted longer or peaked well after the storm was over. The problem is that detention ponds are a promise with many conditions. And rarely are these conditions adequately met in the local, on-site situation.

If we take the conditions and impacts of water volume, we find that these cannot be met solely by designing a singular system for a site. Once a storm was over, and all of the detention ponds along a particular stream were letting out their water at a controlled rate, volume increased and flooding recurred. The problem of "peak flow" that many of these systems tried to mitigate was perhaps solved in their immediate vicinity, but downstream, no one was able to see that detention had taken place.

To address stormwater issues at the local level, the federal government felt obliged to intervene. With the introduction of the National Urban Runoff Program in 1986 and the passage of the Water Quality Act in 1987, stormwater pollution and mitigation became part of the national agenda. Stormwater-quality regulations demanded that local governments deal with water contamination. Many cities and counties spent large sums of money on untested

structural best-management practices (BMPs) and unworkable regulatory controls, trying to meet ill-defined standards of "maximum extent practicable."
Many of these BMPs were nothing more than pollutant traps, with failure rates
as high as 80 percent for infiltration basins.[3]

Furthermore, the result of these types of ordinances has been more often than not the creation of ugly depressions lined with rocks and invasive
grasses. Because many of these ponds are located in urban areas near residential structures, a detention pond designed to store water temporarily must be
entirely fenced in, with a barrier at least 3.5 feet high. Such measures to avoid
possible risks, minute as they may be, illustrate the inadequacy of rules that,
while addressing one problem, create another. (See figure 6.1.)

Measuring Health

Criteria for establishing sound stormwater programs are not met by applying
mechanical engineering standards of flow and volume. Rather they should be
based on ecological biocriteria for a particular region and locale. Under this
paradigm, a stream restoration or conservation target is a measure of biological health, and the stormwater program is focused on how to attain and maintain this health. This concept is already leading some agencies to develop
bioassessment procedures, which are modified for different states or regions.
These procedures can be carried by volunteers, who monitor existing conditions through simple test measures.

Rather than following strict rules and formulas for controlling volume and flow, we must change the way we design developed places. Taking cues from ecologists and biologists, designers could mimic nature when appropriate. In other places, it may be that mechanized and technological solutions are the answers. Easily achievable, if allowed, are changes to the physical design of streets, parking lots, and sidewalks. There is no doubt that narrowing street pavements, using shorter and shared driveways, reducing cul-de-sac radii with vegetated islands, utilizing porous materials, and creating infiltration areas could be implemented in most of our suburban and urban locations.

Finally, both the public and public officials should understand that temporary standing water is not indicative of a failed system. Despite misconceptions created by the fast water-removal approach, the public should understand that temporary standing water may be a sign of efficiency and good ecology. Such water is being held back to recharge depleted groundwater and prevent downstream flooding, while maintaining the integrity of a whole system. Overall, stormwater needs to be regarded as a resource and not as a waste product that must be removed as soon as possible.

Small-Scale Site-Distributed Approach

The backbone of the stormwater paradigm change is the shifting from rigid formulas and water-measurement standards to a holistic, small-scale, distributed, and accumulative approach. At the heart of this shift is an infrastructure metamorphosis. What must be impacted is the overall stream health and water quality. And this can be achieved by changing the way we deal with the flow of water. Conveyance systems, which typically comprise curbs and gutters, inlet and outlet structures, and piping systems that move water from source areas to centralized control areas, should be reconsidered. Instead, infiltration systems and nonstructural conveyors that aim to mimic natural hydrologic cycles can be used.

The design and implementation of such healthier ecological approaches is quite dependent on the permit process. Since most jurisdictions still follow the outdated practice of structural conveyance systems with wet and dry detention ponds, changes may occur through economic incentives.

Some developers and residents are realizing the shortcomings of the traditional detention ponds. Increasingly they are seen as expensive to design, construct, and maintain. It is often estimated that stormwater ponds in new, suburban developments consume approximately 10 percent of a project's developed land area.[4] Other research shows that the cost of a conventional conveyance system typically ranges between $40 and $50 per linear foot. With such costs, the elimination of one mile of curb and gutter can decrease infrastructure and storm-conveyance expenditure by approximately $230,000.[5]

The desire to change practices at the local level has created a momentum that may slowly shift the design paradigm and alter the outdated and inflexible rulebook. Groundbreaking residential projects such as The Woodlands, Texas, and Village Homes in Davis, California, demonstrated as early as the 1970s that conventional suburbia can be designed differently.

Both developments used well-drained, gravelly soil to locate areas for percolation ponds. Both can be described as miniature watersheds with dendritic systems of swales, and shallow ponds designed to handle and absorb storm events entirely on-site, including streets. Yet both of these pioneering developments, with their grassy drainage swales, curbless narrow streets, and constructed wetlands, have failed to generate wide legislative appeal—they were ahead of their time.

It took almost 25 years for local public officials to initiate programmatic transformation. In the 1990s, Prince George's County, Maryland, pioneered several new tools and practices to handle stormwater runoff. Encouraging developers to participate in the process, and input their ideas, the county was willing to accept various solutions as long as it attained the overall goal of imitating predevelopment hydrologic conditions. Indeed, in one example, a developer saved nearly $300,000 when the use of individual-lot bioretention practices alleviated the typical solution for a pond, and the area was converted into six extra lots.[6]

Another case of local innovation is Seattle's pilot Street Edge Alternatives (SEA Street). Designed to provide drainage that more closely mimics the natural landscape prior to development than traditional piped systems do, the program reduces the impervious surfaces of existing streets. By narrowing streets, eliminating curbs and gutters, introducing vegetative swales, and adding trees, the program reduced impervious surfaces to 11 percent less than a traditional street.[7]

The method used for achieving this goal was to maximize the stormwater time of concentration and the site's detention volume, without compromising the homeowners' access and parking needs on the street. Two years of monitoring show that the SEA Street program has reduced the total volume of stormwater leaving the street by 98 percent for a 2-year storm event. Furthermore, monitoring shows a positive response by residents who see their streets as more pedestrian friendly, soft-edged, and environmentally sound. (See figure 6.2.)

These types of small steps taken by local jurisdictions have had an impact beyond regional borders. Strategies for the design of low-impact developments, including the creation of alternative stormwater controls, are influencing state and federal agencies. In 2003, the U.S. Department of Housing and Urban Development published its report *The Practice of Low Impact Development.* In it, the agency called on developers and local jurisdictions to

Before

After

Figure 6.2

Seattle's pilot Street Edge Alternatives (SEA Street) is designed to provide drainage that more closely mimics the natural landscape prior to development than traditional piped systems do; the program reduces the impervious surfaces of existing streets. By narrowing streets, eliminating curbs and gutters, introducing vegetative swales, and adding trees, the program reduces impervious surfaces to 11 percent less than a traditional street. (*Source:* Courtesy © City of Seattle) incorporate alternative techniques and technologies, providing ways to simultaneously incorporate economic and environmental considerations into the land-development process.

Ultimately the endorsement and backing of local jurisdictions of unconventional practices together with their economic benefits would create momentum for change. Experimentation and fluidity would replace exacting rules. Principles would transcend codes and standards.

Working the Earth

One of the most troubling stages in the site-development process involves clearing of vegetative cover and mechanical grading. The use and attributes of the heavy equipment, and the desire to cut costs by executing massive grading, often result in complete alteration of the landscape and degraded environmental conditions. Clearing and grading impacts include excessive sediment, greater flow volumes and velocities, and loss of riparian vegetation. The Environmental Protection Agency has estimated that approximately 600 million tons of soil erodes from U.S. construction sites each year.[8] (See figure 6.3.)

Federal agencies and local governments have generally recognized the consequences of clearing and grading practices, and have imposed mandatory standards and regulations. Two of these are the National Pollutant Discharge Elimination System (NPDES) and the Coastal Zone Act Reauthorization Amendments (CZARA), enacted by Congress in 1990 to address the impact of nonpoint source pollution on coastal waters. NPDES is a stormwater permit program enacted by Congress in 1987 under the Clean Water Act (CWA), and is mainly concerned with stormwater discharges associated with industrial activities. CZARA, on the other hand, requires each state with an approved coastal zone management plan to develop a Coastal Nonpoint Pollution Control Program. Many states and local jurisdictions have followed suit and developed their own standards of practice based on these two programs. However, many of these local manuals fail to address the larger issue of erosion and grading. Rarely does a jurisdiction review and deal with grading and earthwork plans devised for larger developments, or examine their impacts. Configuration of streets' easements and their center-line elevations, as well as lots and build-

Figure 6.3
Attempts to reshape de-
velopment are often
thwarted by engineering
standards and proce-
dures. The impacts can
be seen in the administration of street layouts
and widths, as well as
grading and drainage
practices. One of the
most troublesome stages
in the site-development
process involves the
clearing of vegetative
cover and mechanical
standardized grading.
(*Source:* Courtesy ©
Landslides Aerial Pho-
tography)

ings' final-floor elevations, are normally designed and assigned by professional consultants facing little or no scrutiny from public officials. Existing regulations typically resort to requirements for the installation of structural controls such as sediment traps. These types of regulations are not only poorly implemented and enforced; they lack a comprehensive look at the totality of the impacts.[9] Structural control measures primarily control sediment and do not prevent erosion. Most erosion- and sediment-control manuals discuss proper construction of temporary diversion, conveyance, and containment devices, and methods of quick and effective revegetation, but they fail to address the overall approach of massive grading and slope cuts to accommodate the ease of construction.

The mechanical nature of earthwork, and the ever-growing reliance on larger, faster equipment, often precludes delicate site grading. When hundreds of acres are to be prepared for the construction of homes, three major factors drive the grading operation: regularity, maneuverability, and the balancing of cut and fill.

While the overreliance on large types of mechanical apparatus to reshape the earth may have generated the homogeneity of our built landscape, a few new approaches may open the way to a paradigm shift in grading practices. In residential construction, the incorporation of affordable pier foundation systems, rather than concrete slabs, may in the future eliminate unnecessary earthwork. Systems such as the Chance Helical Pier Foundation allow for little or no grading work on a site. The piers are self-leveling, and finished floor height can be raised to whatever level is desired. They require little engineering and construction effort. A two-person crew with a skid loader can place over fifty anchors in less than one day.

While shifting to mass production of pier systems in subdivision construction may lie far in the future, other more conventional techniques for improving grading are gaining popularity in the construction industry. These technologies, integrated with global positioning systems (GPS), allow operators of grading equipment to "see" design surfaces, grades, and alignments from inside their cab, and in some instances even projected on their windshield. Displayed in the operators' field of view are cut or fill alignments as well as future roads and building sites. (See figure 6.4.)

Figure 6.4
New technologies such as GPS and digital windshield display, provide better management and control of heavy equipment. (*Source:* Courtesy © Trimble International)

The placing of the digital site plan in the cab enables operators to bring the site up to a consistent and accurate grade that makes it possible to execute precisely more detailed grading. "It's almost like a combination of grading and connecting the dots," noted one of the users, adding that "it always shows the operator where he's at as well as how close he is to where he needs to be on alignment."[10]

While such operation still requires the skillful hand of an operator, new developments integrate the site plan with mechanical control. These technologies tie the three-dimensional grading plans into the machine's hydraulics to automate the blade within 1-inch accuracy.

The integration of three-dimensional GPS grade-control systems into machinery such as Caterpillar has already made its mark on the earthmoving industry, as attested by one of its users:

> You can use a laser on a flat building pad or on a road that has, say, a crown of 2%. But it doesn't work on a surface like a golf course that is moving up and down longitudinally. Before GPS machine control you could only set the blade at a certain percentage and grade it that day. You couldn't move up and down as far as height or vertical in the same direction. Take a typical residential subdivision. Before we equipped our machine with 3D machine control, we'd have a surveyor stake it and we'd basically grade by reading the stakes. Then when we got ready to pave the job, we'd take out a laser, set up the transmitter, and do a final grading. With 3D there's no staking, so we can start faster, and if there are any changes, we can do that in the field using the CAD file or design file.

Back to the Site

While the use of attractive new technologies such as three-dimensional GPS or the Pier Foundation system may aid in changing current practice, a simpler solution lies in letting commonsense planning take center stage. At the core of such an approach must be a fundamental understanding by regulating agencies that each site or each locale presents a unique set of conditions that influences

Figure 6.5
Solutions to site development put forward with the intention of advancing current ordinances, rather than mindlessly following exacting standards, must be embraced. Otherwise, the physical and spatial pattern illustrated will prevail. (*Source:* Courtesy © Landslides Aerial Photography)

the design solutions. Some of these solutions may not fall under existing categories of regulations or standards, yet if they are the result of a thorough, accurate site analysis, they may represent a preferable approach.

Insensitive stormwater regulations and mass-executed grading techniques are just two examples of development standards and formulas that do not result in minimal impacts. Such standards may require stormwater-treatment devices inappropriate to site and locale.

Thus we may find the enforcement of detention ponds in dense urban areas, and the construction of pipes and mechanical pollutant traps in open subdivisions. Design of places cannot be done in a mechanical fashion, to "meet the numbers" of codes and regulations. (See figure 6.5.)

Historically, engineers have assumed primary responsibility for identifying a site's resources and integrating them into project designs. These professionals, however, usually have not undergone the specialized training necessary to carry out their assigned tasks in the context of alternative site-specific solutions. At the center of this process must be a collaborative multidisciplinary approach to planning. Solutions must benefit from the input of a variety of natural resource and land-development professionals, including biologists, ecologists, and hydrologists. Regulating agencies should then approach solutions put forward with the intent of advancing current ordinances, rather than resorting to rigid standards and added controls.

Finally, to attempt new solutions we must have the backing of public officials even at the risk of failure. All professionals agonize about their potential liability should design fail. And with the United States being one of the most litigious societies in the world, an overly conservative approach to development is increasingly triumphing over imagination, innovation, and experimentation. Professional and public officials should realize that in many states, under statutory immunities titled "design immunity," a public entity is generally not liable for injuries caused by a dangerous condition of public property if the following three essential elements are satisfied: (1) a causal relationship between the plan or design and the accident, (2) discretionary approval of the plan or design prior to construction or improvement, and (3) substantial evidence supporting the reasonableness of the plan or design. As stated by the courts in several cases, this type of immunity reflects a legislative intent to insulate discretionary planning and design decisions by responsible public officials from review in tort litigation.[11] These acts are particularly important because liability and legal issues are often cited by jurisdictions and professionals as the most critical issues associated with the implementation of unconventional designs and modified standards.

The spread of homeowners associations, condominium associations, and cooperatives is another example of how liability and legal issues are transforming place making and development design. As seen in chapter 5, the survey of municipalities and developers demonstrates the existence of two sets of standards and design parameters: those that pertain to the public domain and those applied to private communities. Public officials often regard the latter, with their privately owned streets and open spaces, as a tool for promoting flexible planning, frequently resulting in innovative and efficient land use and original layouts—characteristics absent from conventional subdivisions. Developers see private communities as a medium for a simplified approval process and the introduction of design innovation. They are using private development to push the density and efficiency envelopes while protecting environmental resources and increasing marketability and financial returns. Public officials agree that because the local government has no legal or maintenance responsibilities for private development, and is thereby freed from liability concerns, such communities often use land more efficiently, through clustering and narrow-street systems. We must recognize that the current practice of allowing different sets of standards for private developments acknowledges the inadequacy of standards applied to public ones, and validates the impression that typical regulations are not determined by actual performance, marketability, or good design.

Part III

Altering Inherited Traits

7

Private Places and Design Innovation

> The building codes of many of our cities are obsolete, drawn to favor certain industrial trades and certain types of merchandise which create unnecessary cost of home construction.
>
> —J. C. Nichols, 1945

The de facto legal and regulatory landscape of the United States has been altered by the vigor and popularity of small managed places.[1] The late twentieth century witnessed record growth of private residential communities. Collectively referred to as common-interest communities (CICs) or common-interest developments (CIDs), these communities rely on covenants, conditions, and restrictions (CC&Rs) to privately govern and control land use, design decisions, services, and social conduct. The communities own, operate, and manage the residential property within their boundaries, including open space, parking, recreational facilities, and streets. Although CIDs have historically been the domain of the affluent, they are now becoming the choice of both suburban and urban residential development. Taking the form of condominiums, cooperatives, and single- and multifamily homes, gated and nongated private communities are spreading, worldwide, across diverse economic and social classes. (See figure 7.1.)

This phenomenon is causing an unprecedented transition from the traditional individual ownership of property to collective governance of most property in the United States. This is a remarkable departure from the individual ownership of property that has been part of the tradition of the American political and economic landscape. The trend, at the very least, establishes a new microscale level of government beneath our municipal structures.

Figure 7.1
Singer Island, Florida,
typifies common-
interest communities.
Almost all developments
seen in this image, from
single-family homes to
high rises, are develop-
ments run by collective
private ownership of res-
idential property and
outdoor space. (*Source:*
Courtesy © CORBIS)

Indeed, the numbers provide a clear indication of this movement's strength. At the end of the twentieth century, about 47 million Americans lived in condominiums or within cooperative and homeowners associations. Growing from 500 neighborhood associations in the 1960s to an estimated 231,000 in 1999, homeowner associations now compose almost 15 percent of the national housing stock, with an estimated addition of 8,000 to 10,000 private developments each year.[2] In the 50 largest metropolitan areas, more than half of all new housing is now built under the governance of neighborhood associations. In California—particularly in the Los Angeles and San Diego metropolitan areas—this figure exceeds 60 percent.[3]

Some authors, such as Barton and Silverman, Blakely and Snyder, Nelson, and McKenzie, suggest that the spread of CICs is driven by the mutual interests of developers, consumers, and local governments (including planning officials).[4] Developers benefit because they can maintain profits—despite the high costs of land and infrastructure—by introducing efficient land-design schemes and, often, higher densities. Consumers, in their increased ability to control their neighborhood character and aesthetics through compliance and enforcement mechanisms, see a way to protect their property value. They also see CICs as providing greater infrastructure provisions, recreational amenities, and community services. Local governments prefer CICs because they privatize infrastructure and reduce public costs. As McKenzie says: "The cities can acquire new property tax payers without having to extend them the full panoply of municipal services and thereby making Common Interest Developments 'cash cows' for local government. Some municipal governments have begun to virtually *require* that new housing construction consist of Common Interest Developments."[5]

The growing fiscal crisis of many local governments often means that they are unable to handle the demands of building and maintaining streets, collecting garbage, and providing other services. In response, the establishment of a separate legal mechanism within a community, such as a neighborhood association, allows collective control over a neighborhood's common environment and the private provision of common services. Perhaps more importantly, this also creates a de facto deregulation of municipal subdivision standards and zoning because cities and towns allow for a different, more flexible set of standards to be implemented in such developments. Often, the results are innovative spatial and architectural layouts, and, sometimes, unusually sensitive environmental design. Such a shift in neighborhood governance enables a resultant shift in the design of residential developments, a shift that heretofore has not been fully appreciated.

How widespread is this phenomenon? What are the attitudes and perceptions of public officials and developers with regard to subdivision regulations and their impact on privately managed and controlled communities? And do some of these developments indeed push the planning envelop to attain desirable design outcomes such as increased densities?

The Dichotomy of Common-Interest Communities

Typically, urban planners and sociologists bemoan the growing popularity of private communities. Davis, in *City of Quartz,* and Garreau, in *Edge City,* lament the replacement of the pluralistic city by spatial segregation.[6] They see this spatial segregation as resulting in the marginalization of exclusive residential, retail, and transportation spaces. Blakely and Snyder, Lang and Danielsen, Stark, and Franzese describe private-community regulations, such as prohibiting pets, limiting how long a garage door may be left open, the amount of grass, trees, and shrubs on a property, and the kind and color of window treatments.[7]

Barton and Silverman claim that common-interest communities fail as participatory democracies because their properties do not reduce but rather intensify conflicts within the community, as people assert their property rights against one another. Furthermore, many of these communities are home to renters who have no voting rights or due representation in decision making about the places they live in.[8]

Private communities, particularly gated ones, are also the target of social critics who see in them an exclusionary and elitist means by which the rich can physically segregate themselves from the lower and middle classes.[9] Some suggest that urban fear drives people to live behind gates: "Gated communities respond to middle-class and upper-middle-class individuals' desire for community and intimacy and facilitate avoidance, separation, and surveillance."[10] On the other hand, some say that such communities also keep the wealthy in the inner city or attract them back to it. They suggest that whereas the neighborhoods themselves may not be integrated, the city as a whole becomes more mixed.[11]

It should also be noted that although CICs are often gated and walled, there are both private communities that are not gated, and public developments, such as public housing, that are. In fact, most CICs—72 percent—do not have any security system in place, while 11 percent have staffed gates or coded-gate systems, and 17 percent have security patrols.[12] Furthermore, the 2001 American Housing Survey (AHS) suggests that the desire for separate living behind gates traverses economic class and race. Analyzing the AHS, Sanchez and Lang show, contrary to the notion that all gated communities are affluent and predominantly white, that there are also gated communities inhabited by minority renters with moderate incomes.[13] According to the data, renters, who are more ethnically diverse and less affluent, are nearly 2.5 times as likely as homeowners to live behind gates or walls, and over 3 times as likely to have controlled entries. Regardless of whether they are renters or owners, Hispanics are more likely to live in such communities than are whites or blacks.[14]

Private CICs are also gaining diversity in types of housing. The 1999 Community Association Institute Survey indicates a general distribution of about 67 percent single-family homes, 15 percent condominiums/apartments, 14 percent townhouses and duplexes, and 2 percent mobile homes. Two-thirds of the associations surveyed have fewer than 500 units; 6 percent have 500 to 999, and another 6 percent have 1,000 or more.[15]

CICs are rapidly being popularized in other parts of the world. Press coverage and research from Europe, Africa, South America, and Asia suggest a global phenomenon. *The Economist,* in a recent article, revealed that former Prime Minister Margaret Thatcher moved into a "gated community" in south London, and that although gated communities are still rare in Britain "many people in Britain like the idea of living somewhere safe, fenced-off and privately guarded."[16] In South Africa, where secure communities were an unavoidable consequence of racism, postapartheid gated private developments are inhabited by all races and not only by the rich.[17] In Saudi Arabia, private compounds of linked houses provide extended families with privacy and identity. Such privately owned compounds seem to be a reaction to the single-residential typology imported from abroad during the country's modernization period.[18] Since the early 1980s and the economic reforms, more and more residential areas in Chinese cities have walled themselves off to improve security and define social status.[19] (See figure 7.2.) Private communities in Southeast Asian countries, such as Indonesia, are marketed as places that allow the differentiation of lifestyle, and give prestige and security to their inhabitants.[20] In Latin America, sprawling private gated communities at the metropolitan edge

Figure 7.2
Private common-interest communities are gaining popularity in other parts of the world. Many of these communities, such as this one in Shekou, ShenZhen, China, are designed by American companies and are based on U.S. planning and design standards. (*Source:* Courtesy © Jodie Misiak)

Figure 7.3
In Latin America, sprawling private gated communities at the metropolitan edge of cities such as this one in Santiago, Chile, have become the norm for a growing sector of the population in search of security and efficient privatized "public" services. (*Source:* Courtesy Gloria Yanez Warner)

of cities such as Santiago and Bogotá have become the norm for a growing professional class in need of a relatively secure lifestyle in an environment dominated by social and economic poverty.[21] In Buenos Aires, the deteriorating political and economic situation has resulted in developers and private companies controlling and providing privatized "public" services. Such services attract large sectors of the population to large private developments, in which half a million people now live.[22] (See figure 7.3.)

Dual Governance—Dual Rules—Dual Design Outcomes

The proliferation of CICs and privately owned and managed residential subdivisions in the United States is also backed up by the results of a nationwide survey of public officials and developers. The survey indicates that in the majority of the jurisdictions surveyed (84 percent or 130 jurisdictions), privately owned subdivisions are allowed to be built. Out of these 130 jurisdictions, 63

Table 7.1 Perception of the design characteristics fostered by private subdivisions

Residential private subdivision characteristics	Percent developers, $n = 80$	Percent planners, $n = 145$
Encourages housing clusters	42	49
Permits greater density	25	26
Permits housing types not found elsewhere	37	41
Allows narrower streets	49	61
Allows innovative design	67	57

(43 percent) have seen the construction of 10 or more private subdivisions in the last 5 years.[23]

Public officials acknowledge the particular design benefits associated with private subdivisions. Fifty-seven percent indicated that in their view such private developments are introducing innovative design in the form of building arrangements and the encouragement of unit clustering. Forty-one percent felt that such developments permit the introduction of housing types not found in other developments in their communities, and 61 percent indicated that they allow for narrower street standards to be incorporated. This perception is relatively persistent in the minds of public officials and developers alike. (See table 7.1.)

While public officials see the benefit of private developments in pushing the design envelop within the confines of the development itself, many are also concerned about the social implications and impacts of these developments on the surrounding communities, as one respondent writes: "As a matter of policy, gated private communities are discouraged as they are not in keeping with the urban form which calls for an interconnecting network of vehicular and pedestrian movement. In addition, the walling of neighborhoods from arterial roadways should be avoided by alternatives such as the placement of other compatible uses along the periphery."

Although almost all of the public officials (82.5 percent) reported that their jurisdictions require private developments to follow established subdivision regulations, the enforcement of these standards through the approval process is malleable. In some cases, when such a development is classified as a condominium, which may include attached and/or detached dwelling units, no formal review of street standards is required. In fact, the majority of public officials surveyed (61 percent) indicated that their jurisdiction allows for narrower streets to be constructed within private developments. As one of the respondents indicated: "Variances are more easily granted within private road systems since the county will not have any maintenance responsibility or liability. A developer for such a community may obtain waivers to reduce and/or eliminate some design/construction requirements (e.g., tighter radii, unusual landscape islands, subbase thickness, pavement thickness, etc.). The local jurisdiction is willing to grant some of the requested waivers as the ownership/ maintenance responsibility for the improvements will not be the dedicated obligation of the jurisdiction."

The practice of building narrower roadways and offering smaller building setbacks in private subdivisions has become widely accepted in the last decade. A street-standards survey completed in 1995 showed that 84 percent of the cities polled allowed for different street standards in such developments, and that they more readily accepted the introduction of different paving materials, changes in street configurations, and the employment of traffic-calming devices.[24]

As amplified by the survey, common subdivision regulations often restrict alternative solutions. Developers and public officials see in privatizing subdivisions a vehicle for a simplified approval process and the introduction of design innovation.

Such attitudes can also be seen in developers' responses regarding stormwater-mitigation requirements. The majority consider the system layout and the fees associated with hookups excessive. With jurisdictions requiring developers to provide large stormwater-management systems, and with the regulations for these systems following the outdated and expensive practice of structural conveyance (piping), with wet and dry detention ponds, developers may be realizing the shortcomings of existing requirements. Yet while narrowing streets, using alternative paving materials to reduce impervious surfaces, and constructing vegetative swales instead of concrete gutters would reduce costs and create favorable ecological conditions, they are not easily approved.[25] As one of the developers remarked: "Regular subdivision codes don't allow flexibility. Lots are too standardized and streets use too much area. If I could build narrow streets and small lots, developments controlled by covenants and HOA will not be necessary."

As the survey shows, the ability to provide design choice, efficient layouts, and the avoidance of a lengthy approval process drives both sectors to offer CICs rather than typical subdivisions. It is unfortunate that under such circumstances, change is less likely to happen in the public realm through traditional means than at the hands of outliers and renegades. Indeed, it seems that in the last decade most innovation in subdivision design has sprung from within the private domain and under the governance of community associations. The following section describes such cases.

Density, Streets, and Nature

One interesting tool for increasing densities has been the introduction of a condominium form of ownership to single-family developments. Although not new, and based on condominium-enabling legislation of the mid-1960s, the application of such a legal structure to a single-family development is more recent. At the Sancerre development in Newport Beach, California, the developer planned to build and push a single-family residential development to a net density of 9.4 dwelling units per acre. As part of the master-planned Newport Ridge project, the Sancerre site was subject to development parameters established by the Irvine Company; housing on the site was to be geared to a moderate-income market segment. The site itself, which was zoned for planned development, could have satisfied this requirement with townhomes at about 14 units per acre. However, the developer envisioned a stronger market for the site in single-family housing and instead adopted a cluster concept. Under the local planning regulations, such density modifications could not be allowed with the current single-family subdivision ordinance.

By opting for a private subdivision, the developer was able to lay out four- to six-unit clusters around private drives with zero lot lines to maximize the usable open space. Furthermore, the developer determined that the legal framework for condominium ownership would be advantageous. The advantage to the developer was that the project could be developed under multifamily standards, which are geared toward higher-density housing. Standards for street widths, parking ratios, and setbacks for multifamily projects are less expansive than for standard single-family projects. (See figure 7.4.)

Figure 7.4
By opting for condominium ownership, the developer of Sancerre in Newport Beach, California (horseshoe pattern at center), was able to increase the density of single-family housing to 9.4 units per acre. Taking advantage of the county's multifamily standards, the development, situated in the midst of other private common-interest communities, shows higher efficiency of land use and provides more affordable single-family housing units. (*Source:* Courtesy © Globe-Xplorer Airphoto)

As condominiums, each four- or six-unit cluster could be developed as a single lot, where the buyer would get title to the house and to the side and rear yards as defined by the condominium plan. The concept is similar to townhouse ownership—an idea buyers readily understand and accept—but it is seldom applied to single-family units.[26] The result is a small-lot, zero-lot-line, courtyard community that pushes the single-family home densities to the limit in order to meet market demand for this type of housing. Dwelling units are approximately 35 feet wide and most units have a 10- to 15-foot-wide yard on one side, which wraps around the unit and flows into a 15-foot-wide rear yard. The design flexibility inherent in the cluster layout leaves room for integrating the open land into and around the groupings of structures. This ensures ready access to considerably more open land than would be possible with a conventional pattern.

As seen in the survey, most jurisdictions allow for a different, more flexible, and narrower set of standards to be implemented on private streets. A case in point is the developments of Belmont, Virginia. The Belmont plan originally incorporated a curvilinear-loop street system that conformed to the Virginia Department of Transportation's (VDOT) subdivision street requirements. With Loudon County adopting a neotraditional neighborhood-design initiative, design features such as narrow lanes, reduced rights-of-way, and smaller turning radii were sought by the developer. However, to achieve these changes at least eighteen variances from VDOT's Subdivision Street Requirements had to be requested.

After prolonged and unsuccessful negotiations with VDOT, these requests were withdrawn and, with the cooperation of the county, a private local street system to be maintained by a homeowners association was developed. Only three collector and arterial streets were designed according to the prevailing subdivision standards and accepted into the public street system; the rest of the street system was placed in private hands. (See figure 7.5.)

Belmont was the first development approved by the county that incorporates an extensive private street system. Although the original plan was never

Figure 7.5
Belmont, Virginia, was the first development approved by the county that incorporates an extensive private street system. Although the original plan was never implemented due to financial difficulties, it subsequently forced the county to adopt its own set of private street standards to be used on future projects. By 2003, two new CICs, Belmont Green and Belmont Forest, were constructed based on Belmont's original principles: gridded, connected narrow street systems under the control and maintenance of a homeowners association. (*Source:* Courtesy © Globe-Xplorer Airphoto)

implemented due to financial difficulties, it subsequently forced the county to adopt its own set of private street standards to be used on future projects.[27] By 2003 two new CICs, Belmont Green and Belmont Forest, were constructed along Belmont's original principles and under the control and maintenance of a HOA.

Similar situations can be found in other localities. The designers for Southern Village in North Carolina planned on introducing neotraditional design elements such as reduced lot lines and garages accessible through back alleys. However, the town's engineering and subdivision standards conflicted with the dimensions desired by the developer. For example, while Southern Village originally planned its alleys to be 12 feet wide, the standards called for 20 feet of width. Insisting on the 12-foot dimension, the developer and the town compromised by designating the alleyways as privately owned therefore eliminating the town's obligation to provide public services such as trash removal.

The concept of private communities as environmentally sensitive developments may sound to some like a contradiction in terms. However, some of these developments provide examples of responsible construction that minimizes environmental impact while maximizing economic value.

Dewees Island and Spring Island, for instance, are private subdivisions on the coast of South Carolina. Both communities have established architectural and environmental design guidelines that state their desire to "have as their objective, harmonious integration of the built environment with the island's native environment. To maintain and enhance the island's integrity, to preserve the ecosystems, and indigenous landscape and to reduce dependence on non renewable resources."[28]

Both of these developments went through several unsuccessful starts. Whether stalled by entitlement obstacles, community opposition, and financial difficulties, or, as in the case of Spring Island, by community objections to the approved 5,500 dwelling units and two golf courses on the island's 3,000 acres, both downzoned to lower densities to accommodate environmentally sensitive solutions. Both projects look superior when compared to other developments in the area because the developers not only agreed to protect sensitive natural habitats, but they often exceeded established minimum environmental requirements. Working with environmental groups throughout the development process, the developers established a high performance threshold for the projects such as sewage-treatment systems and environmentally responsible building materials.

Dewees Island and Spring Island have an extensive private-development approval process. They require the engagement of registered architects and landscape architects, site surveys, analysis and site-planning reviews by the respective habitat review boards, preliminary design reviews, final designs, and construction documents. Furthermore, the architect on each project is required to submit a written evaluation of "the way in which the project is environmentally sustainable."[29]

Spring Island, for example, has a Habitat Review Board that makes recommendations as well as reviewing and approving architectural and landscape plans in accordance with the Habitat Review Guidelines and the recorded Declaration of Covenants & Restrictions for the Island. While the guidelines include elements of typical zoning ordinances such as general setbacks, siting, and massing standards, they also address development on a microscale. For instance, each lot on Spring Island has its own specific property setback lines and building envelope. The acceptable size for a Spring Island home is dictated by the site conditions such as landform, tree cover, setbacks, views from neighboring lots, and distance from adjoining lot lines. The Habitat Review Guidelines also make recommendations regarding environmentally sensitive design opportunities and stress site responsiveness in grading and earthwork, natural ventilation, massing, and design quality. (See figure 7.6.)

On Dewees Island every building must meet specific needs in eight basic categories: energy, air quality, water, resource recycling, suitable habitat, communication, transportation, and aesthetics. To meet these requirements six sets of standards are required: efficient resource use, minimal toxicity of materials, preservation of natural systems, restoration of natural systems, quality of community, and economic vitality. To minimize habitat disturbance and stormwater runoff, impervious landscape surfaces are prohibited and crushed limestone roads serve the island's only allowed mode of motorized transportation—electric vehicles. (See figure 7.7.)

The extensive use of pervious surfaces, and the island's closed-loop sewage system, which eliminates all discharge, allow for uncontaminated groundwater replenishment. These environmentally sensitive design features and energy-saving building techniques and materials have resulted in lower power consumption

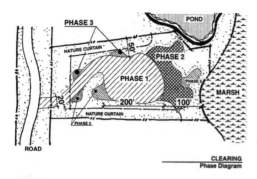

PHASE 3

POND

NATURE CURTAIN

PHASE 2

PHASE 1

MARSH

PHASE 3

200'

100'

NATURE CURTAIN

PHASE 3

ROAD

CLEARING
Phase Diagram

Phase #1- This area represents the initial phase of clearing. The clearing of underbrush can begin by careful bush hogging of a single, narrow and winding entrance connection from the existing road. This swath can be one or two widths of the bush hog and should be aligned in such a way as to respond to any significant specimen trees or other masses of existing specimen vegetation. The intent is to maintain a privacy screen, or "Nature Curtain", from the road right-of-way to the building envelope. The building envelope area can then be cleared in a zone roughly defined by the setback lines on the sides and extending about 200' into the lot from the 100' wide marsh, pond or golf course setback. Although this can be done with the bush hog, careful consideration should be given to the preservation of desirable understory plants to remain such as dogwoods, redbuds, American hollies, tea-olives or sparkleberry. These plants should be identified and hand clearing should occur around these plants.

N. VISTAS

Selective clearing and trimming to open key locations and to frame views from the home is permitted with written approval by the HRB. The objective is to give views from the main rooms of the house while maintaining a filtered view of the house from the river, marsh, pond, trails or golf course. Vista pruning, upon HRB approval, and should wait until after the house has been framed. Clear cutting is not allowed. See also Section 5,B. Unapproved Tree & Shrub Removal or Damage for penalties.

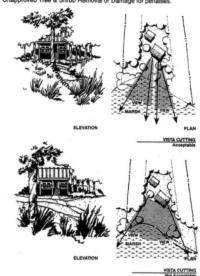

ELEVATION

VIEW
MARSH

VIEW

PLAN

VISTA CUTTING
Acceptable

ELEVATION

VIEW
MARSH

VIEW

PLAN

VISTA CUTTING
Not Acceptable

Figure 7.6

The Habitat Review Guidelines of Spring Island, South Carolina, are administered by an architectural and habitat review board. This board scrutinizes all home plans, including siting, landscape, and any alteration of natural vegetation. The guidelines ensure that each proposed residence conforms to strict environmental standards. Further, buyers are required to spend a day with a local naturalist and attend a two-day seminar on designing with nature before they plan their new home. (*Source:* Courtesy Spring Island Trust)

at 75 percent below the regional average, and a household water consumption of about 30 percent of the regional average.[30]

Although the self-imposed requirements for these developments are much greater than those specified for any typical subdivision, both projects experienced decreased development costs for infrastructure and amenities. Since the existing natural environment was the primary amenity the community was selling, roads were made narrower and built from material previously existing on the properties. Not having to provide paved impervious surfaces enabled the developer to decrease the initial up-front development costs.[31]

Figure 7.7
There are few impervious surfaces on Dewees Island, South Carolina, allowing full restoration of the underground aquifer. Homes are designed to "nest" within the habitat, taking advantage of winter/summer sun, shade, prevailing breezes, and natural lighting. Recycled and nontoxic building products are utilized, and homes are required to use water-conservation fixtures, reducing water consumption by 60 percent. Only indigenous or native vegetation to the South Carolina coastal plains are allowed. This landscaping approach eliminates the need for irrigation, fertilizers, and pesticides. (*Source:* © Courtesy Dewees Island and Jeff Rapson)

Inevitably, the success of these developments is strongly related to the high pricing structure and the narrow market segment these projects have targeted. Both of the developers were able to provide fewer units in their projects as a result of identifying an emerging segment of consumer demand, where the consumer's willingness to pay more enabled a low-density, environmentally sensitive development.

Yet, these high-end communities provide a prototype for other, more moderate projects. The Dewees Island approach has gained the attention of Habitat for Humanity and has become a resource for sustainable design techniques and resource-efficient practices.[32] Other core ideas from both Dewees and Spring Island are serving as models for other lower-end projects such as the 5,000-unit CIC of Palmetto Bluff, Bluffton, South Carolina.

Toward Responsive Design Outcomes

The proliferation of CICs, with their ability to plan, design, and govern outside of public boundaries, can be seen as an indicator of a failed public system.

When developers and public officials resort to privatization in order to achieve a more responsive design outcome, and when local jurisdictions acknowledge privatized communities provide a straightforward way to grant variations and innovation, something is wrong with the existing parameters of subdivision codes and regulations.

Indeed, as Seidel's and our study indicate, for the last 25 years the subdivision-approval process has increased in complexity, in the number of agencies involved, in the number of delays, and in the addition of new requirements.[33] While in 1976 almost half of the developers surveyed seldom required regulation relief such as variances, in 2002 more than half required such a process at least half of the time. Most requests for relief in 2002 were for building higher-density single-family areas, more multifamily units, and the creation of varied site plans such as variation in lot sizes. Both developers and public officials acknowledged that the application process for variances and changes in subdivision regulations is lengthy and cumbersome. Therefore it is not surprising that developers see private developments governed by homeowners associations, not only as responding to market demands and trends, but also as a way to introduce planning and design concepts that are often not allowed or are difficult to get authorized under the typical approval process.

As McKenzie asserts, and our survey correlates, CICs are enabling developers to maintain profits and keep the design process relatively open-ended and flexible.[34] The ability to operate outside the regular, common box of subdivision regulations allows the developers to offer various design solutions that fit the local setting, the targeted site, and the prospective consumers. In some cases these can be attractive, high-density yet affordable single-family developments such as Sancerre, and in others low-density, high-end yet ecologically sensitive construction as on Dewees Island.

Paradoxically, while CICs are often controlled and managed by strict covenants and regulations, their initial design is very much outside the mainstream regulatory apparatus. It is precisely for this reason that they prove to be more flexible in their design solutions and more agreeable to developers, consumers, and local governments. How can such flexibility be integrated into the regular planning process? Can subdivision regulations be made more accommodating and less prescriptive? Will such an approach level the playing field and allow for more housing choices and greater design variety in the public domain? Will such changes encourage developers to plan subdivisions endowed with CIC design qualities but without their restrictive covenants and privatized shared spaces? And conversely, can CICs, while exhibiting great variation in architecture and site-design features, be made less controlling in their management policies?

Obviously, the spread of CICs raises many issues that need to be tackled. But none is more important than the realization that public-policy and subdivision regulations must allow and promote a variety of housing styles and developments. Consumers should not be forced into CICs because they are the only developments that offer them a choice and a range of features. CICs should be seen as a catalyst to change in subdivision standards and regulations and as a bridge between public officials and developers. For through the use of CICs, developers are not only able to circumvent existing regulations, bring down development costs, and in some cases produce quite innovative community-design solutions, but also to enable jurisdictions to secure new taxpayers with less public expenditure.

Not all CICs are created equal, and many are far from perfect. But in terms of design efficiency, utilization of space, and the integration of social and environmental amenities, they illustrate the shortcomings of many standards applied to typical subdivisions.

8

Technogenesis and the Onset of Civic Design

Biomass and infomass are intersected, in some effective combination, where physical actions invoke computational processes, and where computational processes manifest themselves physically.

—William Mitchell

In 1939, the General Motors Futurama pavilion was the most popular exhibit at the New York World's Fair. Long lines formed along two serpentine ramps hugging the facade like climbers on a cliff. The building radiated silver, the narrow slit entrance shone bright red, and inside model cities were nested in a landscape of soft shades of green. It was a gigantic, 4-acre model of America to be: suburbs, housing families in small homes; superhighways connecting towering cities; and cities cleared of industry and crowded living conditions, full of parks and parking.

It was an experience never imagined before. A journey into artificial reality where sounds, visions, and three-dimensional models told a story better than words ever could. In the book *World's Fair,* author E. L. Doctorow says: "Finally we got inside. My stomach tightened and my heart beat as we prepared for the exhibit. We ran and took seats, each of us in a chair with high sides and loudspeakers built into them; they faced the same direction and were in a track. The lights went down. Music played and the chairs lurched and began to move sideways. In front of us a whole world lit up, as if we were flying over it, the most fantastic sight I had ever seen, an entire city of the future, with skyscrapers and fourteen-lane highways, real little cars moving on them at different speeds, the center lanes for the higher speeds, the lanes on the edge for the lower."[1] (See figure 8.1.)

Figure 8.1
The 1939 General Motors Futurama pavilion at the New York
World's Fair was a journey into artificial reality, where sounds and
three-dimensional models con-
veyed a vision of a future place.
(*Source:* Courtesy General Motors)

Yet as fascinating and appealing this fantastic voyage was, it was one-sided: a tale where the plot could not be changed and the outcomes could not be manipulated.

If our aim is to change the paradigm of place making, and challenge rigid, standardized practices, we must let the public comprehend the design intentions for physical spaces. The public's difficulty in visualizing the physical ramifications of standards and regulations is a barrier that must be overcome on the road to better design and planning. Putting into use powerful yet readily available computational tools to introduce communities to the variety of choices available will help them visualize the potential effects that these choices produce, and will ultimately diversify the spatial paradigm of development.

Perceptual Simulations

Simple, interactive, and tangible representations that afford visualization and manipulation of otherwise abstract standards should be integrated into the various coding procedures. Computerized three-dimensional visualization can help those unable to conceptualize the spatial consequences of two-dimensional proposals. Comparisons can be made with existing adjacent parcels like complementing building setbacks and site-design styles. Variance requests can be viewed and evaluated graphically as opposed to relying on just a written application.

While planners and designers tend to think more spatially than public administrators, policymakers, and the general public do, all too often planners and designers fail to highlight the spatial implication of proposed policies. This inattention can be, in great part, explained because spatial consequences have not been made salient in presentation materials.[2] To model their creative visions, designers commonly employ various modes of representation from plans and renderings, to physical and digital models. Modes of imaging technologies ranging from two-dimensional maps, charts, and diagrams to computer models allow professionals to explain their designs and planned interventions more clearly than ever before. Yet, even today few platforms exist that allow immediate, real-time, and seamless changes in response to public or professional input. Often several different modes of representation must be utilized within a project in order to convey different kinds of information and aspects of the design. It is this separation between various representative forms that increases the cognitive load on both the designer and the audience, who must draw relationships between dislocated pieces of information.

Vision

We are in need of platforms that would allow the simultaneous understanding of a wide variety of representations, spanning drawn, physical, and digital forms. Imagine the following scenario in the near future: In a brightly lit room a few people are huddling around a glowing tabletop. Several are holding and moving blocks that look like building models. Others are molding and reshaping the surface with their hands. As one gets closer, a three-dimensional projection of a city's neighborhood is revealed, where fluid, ever-changing images and objects respond to the actions made by the users. This is a setting where physical actions invoke computational processes, and where computational processes manifest themselves physically. As the interaction increases, and the level of excitement rises, the nature of the setting is revealed—an interactive deliberation about the future discourse for an urban site. It is not a simple interchange, but one where data input, discussion, manipulation, and delivery are shaped and configured in real time, allowing for new forms of information delivery that are direct and easily understood by professionals as well as laypersons.

Members of the group—perhaps a civil engineer, a landscape architect, and an environmental planner—stand at the table on which is placed a clay model of a site in the landscape. Their task is to plan a new research park and access roads. Using her fingers, the engineer flattens out the side of a hill in the model to represent a graded parking lot and an access road. As she does so, various colors illuminate the surface from green to yellow. The landscape architect points out that the red colors along the newly created road indicate that it exceeds the allowable slope. They try various configurations by working the clay model. As they shape the terrain, projected colors indicate the resulting impacts—blockage of drainage, erosion susceptibility, and undesirable visual exposure. Finally, as a solution is agreed on, the landscape architect performs a scan of the model to calculate the cut and fill. With the changes projected on the clay surface in various colors, they can clearly see both the location and the amount of earth to be moved and excavated. (See figure 8.2.)

The next day the members of the planning group attend a public hearing at city hall. As they enter the room they unroll onto a large table an enlarged map showing the portion of the site that will contain the entrance plaza. They place an architectural model of one of the site's buildings onto the map. A long shadow immediately appears, at the base of the building model, and follows it as it is moved. They bring a second building model to the table and position it on the opposite side of a large fountain from the first building; it too casts an accurate shadow. "Try noon on December 15," requests one of the public participants. The engineer places a simple clock on the map; a glowing "8:00 a.m." appears on the clock's face. The colleague rotates the hour hand around to noon, and as "8:00 a.m." changes to a luminous "12:00 p.m.", the shadows cast by the two models swing around. It is now apparent that in the winter at lunchtime the fountain is entirely shadowed by one of the buildings. They try moving the building and fountain locations, and on doing so can immediately see the responding shadow patterns. The landscape architect positions a third building, near and at an angle to the first. They deposit a wind-generating tool on the table, orienting it toward the northeast (the prevalent wind direction for the part of the city in question). Immediately a graphic representation of the wind, flowing from southwest to northeast, is overlaid on the site; the simulation that creates the visual flow takes into account the building structures present, around which airflow is now clearly being diverted. In fact, it seems that the wind velocity between the two adjacent buildings is quite high. They verify this with a probelike tool, at whose tip the instantaneous speed is shown. Indeed, between the buildings the wind speed hovers at roughly 20 miles per hour. They slightly rotate the third building, and can immediately see more of the wind being diverted to its other side; the flow between the two structures subsides.

Reality

In the early 1960s, citizen groups spoke out against large-scale urban-renewal projects that came about with little public understanding of their associated physical impacts. As a result, some planners and urban designers became interested in developing new types of imagery and visual simulation to better present and understand the proposed changes. The exploration of new representational techniques as planning tools received an official boost in the United States with the passage of the National Environmental Policy Act of 1970. The act required that all large planning and engineering projects be analyzed for their impact, including visual effects, on the existing natural and human-made environment. Professionals were driven to respond with new simulation tools such as those built at the Environmental Simulation Laboratory (ESL) at the University of California at Berkeley. At the ESL, cameras and scale models were used to examine proposed changes to San Francisco's downtown zoning ordinances

and simulate their urban-design consequences such as new building bulk and height, changes in the city's skyline, and the penetration of sunlight into street corridors.[3] (See figure 8.3.)

Until the late 1980s, computer-based urban simulation was prohibitively expensive. While computers were used to help calculate different camera positions and angles for film making, they were not used to create the simulation itself. The development of computer-generated urban modeling was linked to, and dependent on, the concurrent development of computer hardware and computer-aided design (CAD) software.

In the 1960s, interactive computer graphics were primarily used in large automotive and aerospace companies as well as in government agencies, which developed their own software with room-sized mainframes. In the 1970s, the U.S. Navy began development of three-dimensional programs based on simple geometric forms: boxes, cones, cylinders, and so on. With the development of the PC platform in the early 1980s, CAD software started to gain widespread acceptance. AutoDesk, which released its AutoCAD PC platform in 1983, gained recognition as the industry standard. With the introduction of Intel's

386, the use of CAD spread to many more companies and end-users. It particularly gained momentum in 1988 with its first exploratory release of a three-dimensional modeling system. By the early 1990s, the technology for generating entire landscapes by computer was readily available to design and planning professionals. Yet such simulations required time-consuming calculations in order to generate realistic lighting, reflections, and rendering details.[4] Advances in the PC platform, military flight simulation, virtual reality, and the utilization of the World Wide Web as a delivery system opened new possibilities in the late 1990s. Different types of desktop and workstation CAD and geographic information systems (GIS) allowed cheaper, faster, and often readily accessible simulations. They made it possible for different scales of models to be displayed, with much of the visualization shifting from the desktop to the web. Often referred to as Web3d, these programs allow for the delivery of interactive three-dimensional objects and sceneries across the Internet.

MapJunction3D, for example, is a company utilizing web-based mapping systems to combine fast display of maps, aerial photos, and GIS information in three-dimensional format. Its use of the "Boston Atlas" allows users on the Internet to view two-dimensional as well as three-dimensional maps of the city. With a click of the mouse they can view various data associated with the maps, make queries, compare historical maps to contemporary ones, and view it all in three-dimensional form from various heights. Furthermore, with the right three-dimensional Internet viewer, users can fly over an entire location interactively.

Virtual Places on Two-Dimensional Screens

Available programs have also allowed the creation of realistic shapes and surfaces, and have resulted in the creation of various large-scale urban simulations, most notably the City Simulator of the University of California at Los Angeles. This system combines three-dimensional models with aerial photographs and street-level video to create an urban model that can then be used for interactive fly, drive, and walk-through demonstrations. As part of the Virtual Los Angeles and the Virtual World Data Server projects, the UCLA Department of Architecture and Urban Design (AUD) is building a real-time simulation model of the entire Los Angeles basin. This model will cover an area well in excess of 10,000 square miles and will scale from satellite views to street-level

Figure 8.4
The Virtual Los Angeles project is building a real-time virtual reality model of the entire Los Angeles basin. The downtown portion seen above covers over 150 city blocks. The program will allow multiple simulation clients to seamlessly fly, drive, and/or walk through the Virtual L.A. Model simultaneously. (*Source:* Courtesy © Urban Simulation Team at UC Los Angeles)

views accurately enough to allow the signs in the windows of the shops and the graffiti on the walls to be legible. (See figure 8.4.)

Such systems also involve the user more directly in that, as UCLA's Peter Kamnitzer indicates, they

> permit an observer-participant to insert himself into a dynamic, visual model of an urban environment by means of a visual simulation system employing online generation of color projections onto large screens with as much as 360 degrees of vision. By means of controls which direct his speed and the direction, as well as the movement of his eye, the viewer will be able to 'walk,' 'drive' or 'fly' through sequences of existing, modified or totally new urban environments. Such simulated environments would be initiated either by the researcher or by representatives of government and citizen groups participating jointly in urban planning and design activities. Simulations of sound and atmospheric conditions could be added to heighten the sensory experiences of the participants.[5]

Another interesting example of a virtual city can be found in Tokyo, Japan. Unlike Los Angeles, in Tokyo over a dozen applications have been generated. Sponsored by various players, from the central government to telecommunications and utility companies, these models have been used for various three-dimensional planning applications. Asian Air Survey, which is one of the leading data-surveillance companies in Japan, provides a virtual model that covers the entire area of Greater Tokyo. This model, based on aerial photo data and directly linked to GIS, has been used for landscape planning and telecommunication base-station location, and has also helped in disaster simulation.

The application of three-dimensional computing in the design process has been the drive behind the development of the Tokyo model by the Mori Corporation. Their hybrid digital-physical model is unique. Utilizing both a physical model and computer simulation, they scan the model with a multilens digital video camera that reflects the images on large screens on a real-time basis. This allows the company to present various scenarios integrating for the viewer both a physical three-dimensional model of the site and a projected image of possible scenarios.

Fusing Virtual and Tangible

While urban simulations, such as those described previously, have impressively progressed in the last decade, they are still confined to two-dimensional flat interfaces such as screens. As such, they leave much to be desired from both the end-user and the observer perspectives. They often lack the immediate, tangible interaction that one has with graspable physical objects. And they are missing the understanding and manipulability found in a physical-model representation.

The peculiar interface of computing as we know it was probably set up in 1981 with the Xerox Star workstation. Setting the stage for the first-generation graphic user interface (GUI), it established the "desktop metaphor," which simulated an interaction between a working page on a bit-mapped screen, a pointing device (mouse), windows, and icons. It also set several important human-computer interface (HCI) principles: the "seeing and pointing" and "what you see is what you get."[6] The Apple Macintosh and later Microsoft Windows rendered this style of computer-user interface absolute. Still, in the early 1990s, a few researchers continued to call for new computing visions. In an article titled "The Computer for the 21st Century," Mark Weiser presented his vision of "ubiquitous computing," arguing for a different paradigm of HCI that renders computers "transparent" and tailors their interface to each unique task.[7]

One of the areas of research that investigates the integration of the "real world" and computational media is computer-augmented environments or AR.[8] The most common AR approach is the visual overlay of digital information onto real-world imagery with see-through head-mounted (or handheld) display devices or video projections. Several researchers have tried to create AR-based urban planning support systems. The Envisionment and Discovery Collaboratory (EDC), of the University of Colorado at Boulder, focuses on the creation of shared understanding through collaborative design using an augmented table and wall-size screen. By using a horizontal electronic whiteboard, participants work around a table incrementally creating a shared model of the problem. They interact with computer simulations through the movement of physical objects, which are recognized by means of the touch-sensitive projection surface. This placement of the objects becomes the medium through which the stakeholders can collaboratively evaluate and prescribe changes in their efforts to frame and resolve a problem. On a second vertical electronic whiteboard, the information of the problem at hand is relayed for all to see.[9]

BUILT-IT, of the Swiss Federal Institute of Technology and the Technical University at Eindhoven, demonstrated the use of small Lego-like bricks to control the position and orientation of virtual buildings on a large computer screen. Groups of people seated around a table interact with objects in a virtual scene. A plan view of the scene is projected onto the table where object manipulation takes place. A perspective view is simultaneously projected on the wall. The planar interaction with bricks, however, only provides position and rotation information.[10]

At the Massachusetts Institute of Technology's Media Laboratory, researchers have extended the notion of ubiquitous and invisible computing by associating digital information to everyday physical objects and environments. The tangible user interface's (TUI's) distinct approach is in its focus on graspable physical objects for input rather than by enhancing visual devices. Combining these devices with urban planning discourse has produced new tools essential for an understanding of place making.[11]

The Luminous Table, for example, is composed of projectors and cameras hanging above the table, enabling an attached computer to see the changing positions of different physical objects. The system computes a variety of features that are associated with these objects and projects them back onto the table's surface, moving and changing the features as the objects are shifted or manipulated. For instance, models of proposed buildings placed on the table generate projected data such as shadows, ground-wind patterns, reflective glare, and view corridors. These projections are immediately changed and updated as one moves the buildings around on the table. Because of its dynamic nature, the system can also show the movement of shadows across a site as a day progresses in winter, summer, or any other time of the year, or show prevailing winds as they change by seasons, or demonstrate the increase of traffic at rush hour on surrounding streets. Additionally, two or more tables at different locations may be electronically interconnected, enabling individuals to participate in the design or analysis of a three-dimensional project simultaneously over a distance as a group.[12] (See figure 8.5.)

Another system, Illuminating Clay, allows designers to manipulate three-dimensional models of landforms and objects on which visual data is projected as the shape is formed. A three-dimensional laser scanner continuously scans the three-dimensional model and calculates differences in shape. As the clay surface

Figure 8.5
Communication tools such as *Illuminating Clay* integrate digital and physical forms of representations. These types of tools will provide a major discourse in design and planning processes. Their ability to seamlessly integrate digital and tangible data and update information in real time enhances both the design process and communication to the public. (*Source:* Courtesy © Kath Phelan)

changes its shape, various data such as topography, slope, aspect, cut and fill, or travel time is calculated and projected on the surface. A perspective window screen also allows users to explore the clay model from a person's height. The result is a powerful simulation tool that provides access to a full range of computational resources in a manner that is comfortable and intuitive.[13]

The integration of digital and tangible interfaces provided by the Luminous Table and Illuminating Clay systems is unique in the presentation of urban simulation, where generally the activity of viewing physical models and the viewing of animation and computerized simulations are separate. Along with offering a missing link between the palpable and the digital, the promise of these systems may also be in shaping a pluralist planning process. Typically, during public planning reviews, suggestions and input cannot be immediately simulated and explored, and often require repeated meetings and presentations. These tools, on the other hand, allow for a seamless input-output planning or design process. Ideas, changes, and suggestions and their resulting impacts can be seen and explored in real time, allowing the public to be better informed and involved.

Enabling Sense and Place

Urban planning is also about communicative process and educating the public. As Judith Innes has suggested, this is because the planners also

> interact with regulatory and funding agencies as well as with citizens. They make presentations and act as educators about the issues of planning. . . . What planners do most of the time is talk and interact. Researchers on planning practice have demonstrated that this 'talk' is a form of practical, communicative action. That is, dialogue and other forms of communication in and of themselves change people and situations. Research shows that not only is communication central to planning, but that as researchers and practitioners we must give far more explicit and systematic attention to this basic dimension of practice. Planners are deeply engaged in a web of communicative and highly interactive activities which influence public and private actions in direct and indirect ways that have only recently begun to be recognized in planning.[14]

Indeed, communicating design ideas is required throughout the planning process, where the knowledge involved is never the prerogative of any single expert.[15] Communicating ideas to clients, whether they be political decision makers or the wider public, are activities that are key to the success of the process.

An interesting example of the impact of these emerging tools in enhancing a pluralistic planning process can be seen in the application of Virtual Los Angeles to the Westwood Village plan.[16] During the summer of 1996, the Urban Simulation Team was commissioned by a local developer to build a virtual database of a proposed mixed-use development, Westwood Village, and the surrounding contextual neighborhood. This development is located in the heart of the Westwood district of Los Angeles near the UCLA campus, and includes a residential high rise, a multitheater complex, and a central plaza with three levels of high-end retail and restaurant space. The resulting simulation database greatly benefited the development process of this project in that it was used to create an environmental-impact report, and to study the physical impacts of the development plan from various angles and scales. This interactive process allowed the developer to "look" out windows and modify buildings where windows were unexpectedly blocked. The simulation also allowed the development team and the project architect to discover problems in the design of the main plaza's space. In addition, the simulation of the project was used as an interactive visualization tool to inform the Westwood community about the project's physical impacts. The local residents and merchants were invited to a community meeting by the local councilor to experience the proposed development of Westwood Village through the Urban Simulator interface. The participants were able to interactively fly, drive, or walk to any location in the Westwood neighborhood and view the project from that angle. As a result of this consensus-building meeting, the local community was able to give valuable input in the design process and also had certain fears alleviated related to the project's scale.[17]

Engaging the capabilities of such tools, planners and designers become more effective in teaching laypersons the tools of design, translating design solutions into tangible images, amending them to meet the needs of the community, and empowering them to engage in participatory design to build a true consensus.[18] As Councilman Michael Hernandez of the Los Angeles Mayor's

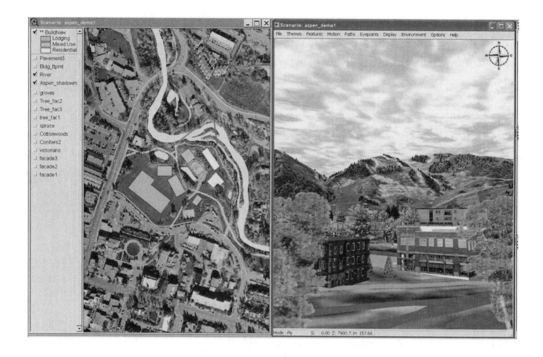

Figure 8.6
Community VIZ software is designed to make the land-use decision-making process more visual, collaborative, and effective by providing capabilities for comparing land-use alternatives. It offers the visualization of numerical computations on geographic data in two- and three-dimensional images in real time to help users visualize the impacts of their alternatives. (*Source:* Courtesy © Community VIZ)

Office explained: "The real value is to allow citizens to participate in the planning process. It really benefits in developing large projects and in looking at the impacts of those projects on communities. With virtual reality, you can actually see it before you build it."[19]

Another interesting example for a place-based decision support and visualization tool is the CommunityViz GIS scenario software. Developed with the support of the Orton Family Foundation, the product lets users create virtual towns and experiment with various scenarios and consequences. (See figure 8.6.)

Like SimCity, the computer game that allows players to build make-believe cities with fictional citizens, CommunityViz is a virtual-reality program. It allows people to envision and see how development proposals might affect their town. More importantly, it enables users to query existing or proposed city zoning maps and then visualize what the community would look like three-dimensionally with photo-realistic buildings, vegetation, and street furnishings that can be modified in real time.

Tools such as CommunityViz have made planning more democratic by promoting informed debate. By showing the likely outcomes of different policy scenarios, interactive three-dimensional tools enhance the decision-making ability of small towns and public officials who are not versed in design. In 2001, Eureka Township in Minnesota, a rural community of 1,490 located 2 miles south of Minneapolis, was one of twenty communities nationwide to receive CommunityViz free of charge. Eureka's proximity to the booming Twin Cities is rapidly changing the character of this rural community. With equipment from the Orton Family Foundation and a $50,000 grant from the Minnesota Office of Environmental Associations, the community group 1000 Friends and the town of Eureka assembled a ten-citizen committee to meet once a month to review and give input on twelve different growth scenarios.[20] "CommunityViz was an important part of our community visioning project," attested Mike Greco, the Citizen Envisioning Task Force chair. He explained:

> Because our small rural township is sparsely populated, it's difficult for our citizens to imagine what dense residential development would look like if it came to our community. The three-dimensional component of CommunityViz allowed us to actually see the potential impacts of development on the landscape. This was a valuable complement to the Scenario Constructor component, which quantified the impact of various growth scenarios on a wide range of indicators, from farmland and natural areas to water quality and the cost of government services. Both of these components of CommunityViz were invaluable to our task force as we compared different hypothetical scenarios for future growth.[21]

The aforementioned innovations in three-dimensional modeling are only the beginning of an ongoing revolution in urban-design tools. As computing power increases and the price of storage declines, the ability to put three-dimensional modeling on the desktop will increase, giving individuals more power to critique proposed changes in their built environment. Putting into use powerful computational tools to introduce communities to the variety of choices available will help them visualize the potential effects these choices produce, and will ultimately diversify the spatial paradigm of place making.

Currently, it is very difficult for the public to clearly envision and understand possible planning scenarios. With new technologies readily available, and others promised for the near future, complex issues can be easily integrated into the planning and design process. With a click of a mouse, end-users can view design configurations and layouts of various developments, density measurements, street widths, and setbacks as well as other related factors.

It is sometimes said that democracy requires an enlightened republic. By promoting the communication that is critical to that enlightenment, new tools that integrate three-dimensional modeling and tangible interfaces will have a profound impact on the politics and discourse of making better places.

9

Places First

In preparing for battle I have always found that plans are useless, but planning is indispensable.
—Dwight D. Eisenhower

As the pressures of metropolitan growth and governance manifest themselves amidst the networks of accumulated regulations, the opportunity for a fresh approach has appeared. In the earliest years of city-building regulations and city planning, the choice lay between the dangerous and costly circumstances of a ruthless private market and public standards and regulations. Now, a century later, the alternatives are either to follow the path of inertia by continuing to pile regulations atop their tottering early-twentieth-century base, or to adopt a flexible set of practices and rules that would allow subdivision and construction to respond to the particulars of place and environment.

Examining the American place-making model, extensively replicated around the world and primarily discussed in this book, leads one to a broader understanding of the impacts of standards and codes on urban development: first, in the global dispersion of uniform formulas and standards; second, in the deficient urban land patterns appearing in the form of sprawl; and third, in the insufficient responses by regulatory agencies despite numerous calls for reform.

With the high rates of growth and the expansion of the metropolitan fringe, concern over the adverse impacts of standard development practices continues to mount. Debate over the nature and type of growth has taken central stage in both the professional and political arenas. Whatever position the debaters have taken, almost all agree that the current forms of land-use regulations and their related codes are archaic and inadequate. At base, the trouble lies in the poor connection between land-use regulation systems and physical design. In the United States, the rationale for regulating development has been based on the separation of uses as devised by the Standard Zoning Enabling Act of the early twentieth century. Unfortunately, another characteristic of the act is that it does not address the needs of physical design beyond rudimentary dimensional requirements. (See figure 9.1.)

The American model also focuses on individual cases and parcels, and neglects to address broader contextual issues of the natural environment, infrastructure and transportation systems, and social segregation. It fosters lot-by-lot rules, seeking uniformity by the application of generic dimensional standards such as building footprints and building setbacks. The deficiencies of this model can be seen in the constant need to amend it. Through the years, a variety of approaches were overlaid on the existing system in order to fit changing goals. Multiple examples can be found in the establishment of overlay districts, historic preservation ordinances, planned-unit developments, neighborhood conservation districts, unified development ordinances, and traditional neighborhood development codes. This type of "band-aiding" has resulted not only in additional layers of regulations, but also in unnecessary levels of complexity because the underlying approach remains unchanged.[1]

Calls for other types of regulatory reforms have been as numerous as the various types of overlay zoning layers and ordinances. Critiques of standards that affect housing affordability, for example, dominate the discussions and the calls for change. As discussed previously, the findings from numerous federal commissions, state committees, and private studies indicate that the typical regulatory envelope discourages efficiency and increases housing costs.

Challenges to regulatory barriers to affordable housing have not escaped the international arena. The World Bank and other lending institutions have attempted to reform housing standards. Policy critics and planning professionals have repeatedly challenged existing practices where countries struggle with standards that were a part of their colonial legacy, or standards that are imported at face value from the industrialized nations. These authors conclude that the key to solving the problem of urban shelters in developing countries lies in the relaxation of existing standards and regulations. They show that in many instances existing standards often impair livelihood by not allowing the

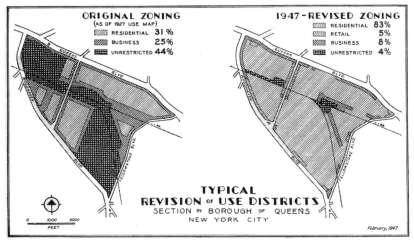

Figure 9.1
With the introduction and spread of "planning by zoning" in the early twentieth century, responsiveness to context, complexity, and change was replaced by an instrument of statistical control. Originally inspired by the need to avoid noxious adjacencies, zoning became widely used because of its simplicity, predictability, and manageability, not because it produced livable neighborhoods or attractive places. (*Source:* New York City Zoning)

incorporation of alternative building materials such as building with soil, or not allowing for incremental construction. According to a recent study, less than half of the urban population in developing countries can afford to build according to the prevailing standards and codes.[2] (See figure 9.2.)

Regardless of the numerous calls for regulatory reform, changes have been slow, at best. A prime barrier to reform, especially in the United States, lies in the merger of the perceived interests of local boards and property owners/voters. The local boards see their role as that of protector of their jurisdictions against cut-rate, shoddy practices. The property owners/voters share this defensive posture because they fear that any change in local rules might decrease the value of their property. For most owners/voters, their residence is their only source of significant capital accumulation.[3]

Attempts to reshape development standards are also thwarted by professional engineering standards and procedures. This obstacle to innovation often appears in the administration of street layouts and widths and their associated grading and drainage practices. Technological alternatives that have been developed for the improvement of waste infrastructure, such as Living Machines, Free Water Surface Systems, and E/One Sewer systems, have failed to gain wide acceptance because they cannot be approved by codes that adhere to long-established engineering practices rather than to performance-based criteria.

Figure 9.2
Many projects in developing countries, such as Trinidad and Tobago, follow codes and standards that were either part of their colonial legacy, or that are imported from industrialized nations as an ideal for modernity and progress. (*Source:* Courtesy © Bob Krist/CORBIS)

The result is that these ecologically appropriate alternatives are rarely advanced as options.

The desire for consistency, particularly in building construction, is understandable. As cities grew and experienced the consequences of disease, fire, and structural collapse, they responded with ever more complex laws. Early in the twentieth century, in the United States, the insurance industry, endeavoring to reduce their losses, developed a model code for states and local governments to enact into law. This model, known as the National Building Code, was first published in 1905. That code was developed by the fire insurance industry and was therefore highly geared to property protection. This code and those that followed (such as the Uniform Building Code of 1927) were heavily influenced by labor-related accidents such as the 1911 Triangle Shirtwaist Factory fire in New York, during which 146 textile workers lost their lives because of inadequate and locked exits.

These early model codes gained widespread popularity because they allowed local governments to adopt technical requirements without the difficulties and expense of place-specific research that a unique code would necessitate. They also guaranteed that the local builders would perform in compliance with insurance standards, so that both builders and the local jurisdiction would be free from personal liability.[4]

Today, building codes and standards, particularly those associated with fire safety, are reasonably enforced through a uniform model. However, the expansion of these universal codes—and the organizations that promote them—into other areas of urban development should not be tolerated, because the result is the global dispersion of uniform formulas leading to standardized place. Unfortunately, code organizations with universal code agendas are continually pressing on. In 1994 the International Code Council (ICC) was established as a nonprofit organization dedicated to the development of a single set of national and international model construction codes, including standardized zoning. By the year 2000, ICC published an impressive array of codes containing an International Residential Code (IRC), an International Private Sewage Disposal Code (IPSDC), an International Property Maintenance Code (IPMC), an International Zoning Code (IZC), and even an International Urban-Wildland Interface Code (IUWIC). By 2003, 33 states in the United States and more than 509 local U.S. jurisdictions had adopted one or another of these codes. Out of these, 32 jurisdictions have embraced the International Zoning Code.[5] (See figure 9.3.)

Model codes may provide an attractive blueprint for communities lacking the financial resources to develop their own codes, but they pose the danger of becoming ubiquitous responses that override the unique needs and possibilities of the particular places that these jurisdictions administer, and whose local policymakers and citizens are endeavoring to protect. (See figure 9.4.)

The Path to Transformation

Although attempts to reform the regulatory envelope have been an integral part of the planning landscape for the last 30 years, the rate of change has been so slow as to reinforce the sense that current standards have attained the power of a generic imperative. Yet, for all their dominance, there is now strong evidence that new modes of governance and new building processes can allow citizens and planners to find better alternatives.

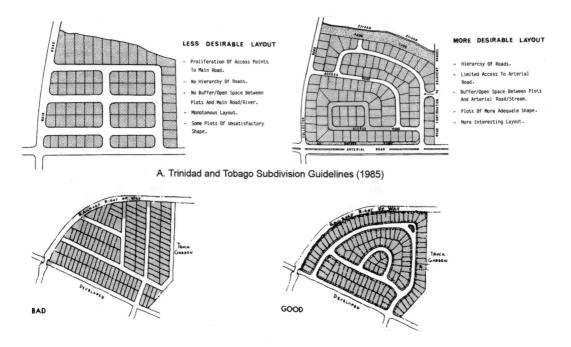

A. Trinidad and Tobago Subdivision Guidelines (1985)

BAD

GOOD

B. United States Federal Housing Administration Subdivision Guidelines (1938)

Figure 9.3
Subdivision standards of Trinidad and Tobago Town Planning Department (A) are taken directly from the U.S. Federal Housing Administration guidelines (B). Standards that are imported at face value from one place to another are the unfortunate results of professional agencies pushing for the establishment of international model codes. (*Sources:* Trinidad and Tobago; FHA)

All the elements necessary for improved planning processes now exist, and all the actors necessary to carry out such reforms are currently on the ground. What circumstances today require is that these elements be joined together in new ways. The new linkage can be achieved by professionals shifting from reliance on standards toward designing and building for the particulars of place. As part of the same effort for improvement, local jurisdictions need to move toward self-determination, master plans, and new forms of citizen involvement.

What follows are recommended initiatives for disentangling the problems inherent in the overindulgence and reliance on standards and codes.

Professional Initiatives

Reinforce the Idea That Design Matters   In the last few decades, decisions regarding the built environment were often made by those far removed from

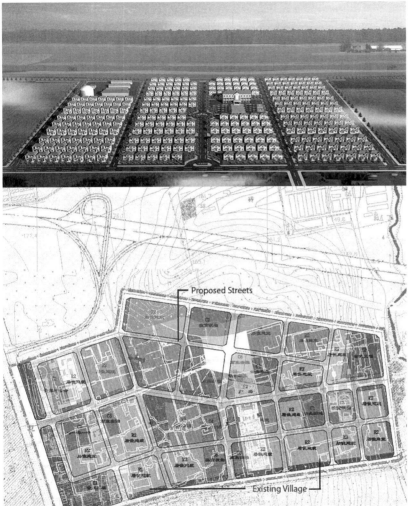

Proposed Streets

Existing Village

Figure 9.4
With an unprecedented
urbanization rate,
China's rural landscape
is being dotted with in-
distinguishable new sub-
divisions (top). In other
instances existing vil-
lages are transformed
into modern develop-
ments by laying out new
street systems based on
international standards
(bottom). (*Sources:*
Henan Province, Bureau
of Construction, and
Heilongjiang Province,
Harbin Urban Planning
Bureau)

understanding urban design and its impacts. The planning professions have generally been reluctant to recognize the importance of physical design, largely because of an ideological commitment to the social science–based disciplines as the foundation for urban planning education and practice.[6] This focus has resulted in the marginalization of urban design and physical planning so that they have all but disappeared from urban planning curricula. The lack of design understanding and an inability to understand three-dimensional consequences leads to an overreliance by urban planners on standards and codes as the instrument of shaping place without questioning either the standards or their outputs. This reliance has not only created a one-dimensional approach to planning, but it has also rendered planning practice inadequately prepared to deal with current ever-evolving environmental and development trends.

The increasing prominence of ecology, sustainability, and culture has brought to the fore the importance of how to physically plan and design the built environment. The core of urban design has shifted from the production of structures; it now draws urban inspirations from the relevant social context, the cultural story of the geographic past, and the natural conditions of sites. As Larry Vale has suggested: "The main reason for a resurgence in interest in physical planning has been the ability of urban designers to link the issues of design to a variety of the most pressing issues of development and implementation."[7] Questions such as how communities should be organized and planned to minimize their ecological footprints and impacts have gained a renewed importance in regional planning efforts.[8]

Such physical-design issues underlie the new-urbanism movement that is now considered a major force in urban planning practice and education.[9] In 2003 the American Planning Association formed a New Urbanism Division. The building industry has also been quick to respond to some of the new-urbanism criteria, declaring that "one of the hottest trends in real estate is the development of town centers and urban villages that include a mix of uses in a pedestrian friendly setting."[10] It is also important to note the growing popularity of physical planning and new urbanism among students.[11] Students are now a driving force behind the resurgence of and emphasis on design and physical planning. This can be seen not only in students' stated enrollment objectives, but also in the establishment of student organizations such as Students for New Urbanism.[12]

This emphasis on physical planning has exposed the inadequacies of practiced regulatory mechanisms. It has also increased the involvement of experts who understand the spatial implications of development in the planning process. As Alex Krieger has observed: "Andres Duany and his New Urbanist colleagues have made us realize the power of *illustrating* the effects of planning: illustrating both good planning principles, in their terms, as well as the consequences of failed policies or techniques such as Euclidian zoning."[13] This renewed bond between design and planning, strengthening the connections between shaping space and its context, and between the expert and the community, is opening doors of opportunity. Going forward, there are the opportunities to challenge existing regulatory practices based on their poor performance, to provide place-based criteria that are responsive to the local and not the universal, to streamline an exhaustive process, and to present a clear vision that a community can grasp, rather than relying on and settling for a "same-based" place obscured by inappropriate rules.

Advance the Role of the Planner and the Designer in Clarifying Codes   As planners defer the drafting of regulations and codes to lawyers, the results are nothing but a complex array of exceptions, qualifications, and incomprehensible language. According to Baer, "Lawyers see regulations not as something to be designed but as a *revelation* of legislative intent. Lawyers do not think of themselves practicing the design of regulations; rather they see themselves as engaging in *exegesis* of legislative committee statements and reports."[14]

As Garvin writes, "There is no need for any intelligent person to face zoning resolutions such as that of New York City, which is replete with language like: 'However, no existing use shall be deemed non-conforming, nor shall non-conforming be deemed to exist solely because of . . . (c) the existence of conditions in violation of the provisions of either Sections 32-41 and 32-42, relating to Supplementary Use Regulations, or sections 32-51 and 32-52 relating to Special Provision Applying along District Boundaries, or Sections 42-41, 42-42, 42-44 and 42-45 relating to Supplementary Use Regulations and Special Provision Applying along District Boundaries."[15] Such language is not limited to the regulations themselves but also extends to codes and standards. Consider for example that the 756 pages of the *International Building Code* of 2000 is interpreted by a larger handbook of commentary in two volumes with over 1,000 pages in order to clarify the intent and rationale of the code provision itself.

Much of the difficulty with regulations lies in the problems of using universal verbal statements to describe particular spatial forms. Codes and standards must be accompanied by illustrations and other visual aids, particularly photographs and three-dimensional illustrations. For such elucidation, they must be placed in the domain of professionals who understand and comprehend the terminology of spatial consequences. While code formulation itself must engage those who can understand three-dimensional design—namely, architects, landscape architects, and urban designers—lawyers, who are essential to the process, must also reform by using common, easily understood language. Finally, academic planning must also renew its commitment to educating other sectors within the profession, such as those concentrating on land use, public policy, and economic development, about the province of design, so that these parties will have the ability to understand the spatial implications of their decisions.[16]

Consult with Professionals Outside the Paradigm The premise of a second opinion with regard to illness is based on the fact that the practice of medicine is not a perfect science. There are many ways to diagnose a particular disease and an assortment of treatments to choose from. It is widely accepted that a second opinion does not vilify the first doctor—it merely allows the patient to get a different viewpoint and opinion.

Like medicine, planning and engineering are not exact sciences. A second opinion, particularly about standards and practices, should be part of the routine. However, unlike medicine, for this process to succeed, opinions should be gathered from those professions that operate *outside* of the specific paradigm in question: for traffic, consult a psychologist or a pollster, for electrical codes a physicist, for site planning an agronomist or a forester, for a sanitary system a limnologist, and so forth. Architects, civil engineers, and planners depend on the continued viability of a professional paradigm and the technological system it is embedded in for their daily work. As long as the current methods are able to meet legal standards and codes, there is little reason to abandon the paradigm or consider radical innovation. But decision makers need to be prepared to consult with people who are not committed to the existing rules; they should seek expert advice from those who operate in professional proximity in order to realize all viable alternatives.

Adopt Local Suitability Criteria   Concerns over growth management and sprawl are prompting organizations to take a fresh look at their current models. At the national level, several professional associations have endorsed local adjustment of fixed national standards. The Institute of Transportation Engineers (ITE), for example, has gone through a reexamination of its street standards and recently even endorsed design practices that are not rooted in prescriptive numerical specifications.

In its 1999 Traditional Neighborhood Development Street Design Guidelines, the ITE, instead of using dimensioning charts and specific design criteria, explains concepts and their underlying logic. The guidelines do not specify a required street width or the number of travel lanes, but emphasize that "a street should be no wider than the minimum width needed to accommodate the usual vehicular mix that street will serve." The document continues by emphasizing that "this simple statement means that a particular traveled surface may be as narrow as twelve, ten, or fewer feet in width. In other cases, streets may be as broad as sixty or more feet. If the principles of design and the balance of these guidelines are read and properly applied, appropriate dimensions will follow as a normal part of the design process for the street under consideration."[17]

It is refreshing to find such flexibility coming from an engineering discipline that often overrelies on prescriptive dimensions. The support and distribution of such a document will allow for variety in local, place-based street design that can only enhance this essential public domain and cater less to automobile use.

Another example is that the American Planning Association (APA), in a major effort to provide new direction, has recently published its *Growing Smart Legislative Guidebook: Model Statutes for Planning and the Management of Change* (2002). Its executive director acknowledges that "it's time we develop new and more flexible codes that can serve all citizens far more effectively than their 20th century predecessors," and the guidebook is the latest attempt to reform planning and land-use laws by providing a variety of options for statutory reform instead of the current one-size-fits-all model.[18]

Professionals should go farther: they can take the lead in proposing ranking systems that can be used to ascertain the suitability of standards by locale. Instead of being prohibitive by setting maximum standards or encouraging the lowest common denominator by setting minimum ones, state and national codes should provide checklists and information to guide development toward an ideal situation. Such future-oriented checklists can make the coding process sensitive to emerging attitudes and technologies for the natural environment. And such checklists can also provide benchmarks for project evaluation, and can be utilized to rank proposed projects according to their suitability in accordance with a set of locally derived criteria.

Adopt a "User-Affordability" Approach in Setting Standards and Encourage Low-Cost Ecological Solutions   In the international sphere setting standards at levels the local population can afford is essential. One of the key problems in many of these settings follows from the habits of engineers who often pursue "modern" performance. Their strong belief in modernity, and their robust design solutions—often chosen because of fears of later poor maintenance—cause them to reject minimum adequacy as an objective. These professionals also often resist variations in service quality by locality, and foreclose designs based on greater affordability. But recognizing the varying ability of consumers to pay for necessary goods like sewer systems, water, and housing can begin the process of improvement and will result in a wider facilitation and acceptance of minimal standards.

Leading educational and lending institutions need to advocate the mixing of suitable modern technology with less expensive local materials and traditions. A move in this direction can be seen in the recent cooperation between the Cities Alliance, the World Bank, and the Massachusetts Institute of Technology to form a resource showcasing and evaluating the experiences gained from upgrading projects in poor urban communities.[19]

Development of both site and dwelling should adopt basic ecological engineering solutions to reduce the use of high-energy systems, the misuse of expensive resources, and the dependency on conventional codes and standards. Using locally recycled, nontoxic materials with low embodied energy as building materials, passive solar and climate-responsive designs, solar hot water, power from wind and photovoltaics, heating, cooling, and humidity control by using breezes, sunlight, and vegetation, on-site use of stormwater, and on-site sewage and graywater treatment reused for irrigation, are some of the steps that can be taken to demonstrate cost reduction, self-sufficiency, and local design compatibility and appropriateness.

Pilot urban and suburban projects such as Christie Walk, a housing development in Adelaide, Australia, and Civano in Tucson, Arizona, provide living models showcasing the suitability of these solutions within the private market.[20]

Establish Best-Practice Clearinghouses   In a climate of increased bureaucracy and complexity, decision-making and legislative changes are slow to occur, but a significant alternative presents itself for both the government and the development professions. The best catalysts for innovation are actual examples. Best practices provide an immediate way to compare experiences and to evaluate projects based on actual performance. They are often the best tools to persuade skeptical decision makers and the public to accept change.

Government agencies as well as international organizations are realizing the importance of such vehicles for innovation. Essential to these clearinghouses is the ability to spread and exchange information over the Internet, disregarding political, economic, and cultural boundaries. Such interactive databases not only allow for the exposure of current and emerging trends, but also promote networking and establish policy development based on what works. The United Nations Habitat, for example, maintains a clearinghouse of best practices for human settlements.[21] This searchable database contains over 1,600 examples from more than 140 countries that demonstrate not only ways to provide improved shelters, but also how to protect the environment and support economic development. It is unfortunate that this powerful tool is currently available only to subscribers and not free for all potential users.

Unlike the subscribers-only UN database, the recently established (December 2002) U.S. Department of Housing and Urban Development Regulatory Barriers Clearinghouse is a free forum to share ideas and solutions for overcoming state and local regulatory barriers to affordable housing.[22] Its services include an electronic newsletter that highlights successful barrier-removal strategies and policies, and a searchable database that offers possible solutions based on actual experiences. Similarly the U.S. Department of Energy's

Smart Communities Network provides a clearinghouse showcasing sustainable-development practices and ecological solutions to construction and development problems. Its site provides model codes and ordinances that communities have used to implement sustainable development, and showcases successful projects.[23]

In an era of media and marketing, the ability to demonstrate achievements and share alternative practices may prove to be the most important tools for change. Planners, and planning organizations in particular, must devote more time and effort in disseminating their experiences and especially, their successes, and make them readily available to others in tangible form.

This group of professional initiatives could move the planning and building professions away from their current insensitive code repetitions, and toward physical designs that respond to the specifics of places and the needs of local and regional environments. Such an advance by the professions, however, cannot succeed unless the local jurisdictions also redirect their attention toward a place-based approach.

Local Initiatives

Expand and Demand More from Local Self-Determination   In the United States, the recent shift in the relationship between national and state governments, and between state and local governments, goes to the heart of the planning process and its decision making. The current movements for "new federalism" and devolution challenge the previous long trend toward international, national, and state uniformity. The new goal, as stated, is to enhance the responsiveness and efficiency of the federal and state systems. More and more responsibilities are passing from the federal government to state and local governments. This process has allocated block grants to states, reduced targeted grants-in-aid from the federal government, increased flexibility for states in complying with federal requirements, and expanded use of referenda and local initiatives by citizens.[24]

The devolution of state governments is also reflected in the relatively slow growth of state financial aid to local governments. With many states facing continuous budget pressures, they have assigned a low priority to helping localities, which leaves these localities to handle their own problems without state interference. Signs of increased local control include the authority to impose local sales taxes and the gradual adoption of limits on the states' abilities to impose unfunded mandates on local governments. Such de facto devolution of state power has been going forward in the United States since 1985.[25]

In recent years, several states have moved beyond de facto devolution and have considered legislation explicitly intended to reduce responsibilities from states for their local governments. For example, some states have considered converting certain existing state aid programs into block grants. Others have combined funding streams for local aid programs to allow for greater local flexibility or have adopted mandates that provide localities with greater control over their budgets, provision of services, and land-use planning decisions.

Devolution, however, brings its own particular limitations and flaws. In its 1991 report *Not in My Back Yard: Removing Barriers to Affordable Housing,* the Advisory Commission on Regulatory Barriers to Affordable Housing identified assorted problems with land-use regulations as practiced at the local level. According to the report, decisions that are made at the local level lack regional and metropolitan considerations. Since they are internally focused on local advantage, such regulations are driving up the cost of housing, making it less affordable and disserving the larger metropolitan public.[26]

Although such findings may seem to call for an increase in federal and state actions, others have argued that the best measures to solve these problems are already emerging at the local level in response. For example, numerous communities have addressed affordability by inventing or adopting tools such as transfer of development rights, cluster and mixed-use zoning, and denser traditional-oriented development. Many of these tools have originated from local initiatives and private consultants, not by federal or state actions.[27]

In its 2002 *Growing Smart Legislative Guidebook,* the American Planning Association (APA) put forward a fresh set of uniform development standards as a planning tool for localities whose goal is mixed-class development. According to the APA, small localities often do not have the resources to adequately modify borrowed standards, let alone formulate new ones. Therefore, they argue, there is a need for a uniform set of development standards that local governments with a small population and/or scarce resources can enact and use to overcome the limitations of earlier codes. According to the APA, "The uniform standards will have undergone a detailed review and analysis by state engineers, attorneys, and other experts that many smaller local governments could not afford, so that the legality and effectiveness of the standards is more predictable."[28]

Unfortunately, the proliferation of such uniform model codes (or model ways to adapt such codes) repeats the failing of the earlier models: they inevitably fail to respond to local conditions. For years now the application of model street and infrastructure standards has resulted in unsatisfactory development forms that waste land and pile on unnecessary expenses. By contrast, the independence of localities and their ability to adapt beyond the central government's yardstick has served as the key component in changing the regulatory envelope. This seems a healthy process that should be encouraged, even at the risk of failed experimentation. Surely the nation's experience with the uniformities that underlie the sprawl of the past 50 years supports such initiatives.

As more communities wrestle with quality-of-life problems due to uncontrolled growth, environmental pollution, and failure of existing infrastructure, they will begin to take a stronger interest in their local power. Such renewed awareness opens the possibility for local communities to establish their own initiatives for localized place-based standards. Under such schemes, local objectives will be translated into measurable limits for air quality, police and fire services, parks and recreation, water, drainage, and traffic. Such initiatives can be undertaken even in the face of limited local financial resources.

If the localities are to realize the best possibilities out of the trend to devolution, they must simplify their regulatory and approval processes, they must redirect their attention toward master plans and vision plans for their communities, and they must explore new modes of professional-government-citizen participation in planning.

Base Standards and Codes on the Level of Local Physical Impact   Local conditions and physical context can provide the threshold for the formulation of development norms. As seen in the case of private development, the style and type of regulations can be place-based, and there can even be emphasis on design details if they are buttressed by explicit approval.

Ideally, in circumstances where a model code is enforced by or adopted from a higher authority, the local jurisdiction must be able to review and amend it to fit its unique situation. A successful example of such an approach can be seen in the adaptation of narrower street standards in Oregon. Although Oregon's land-use laws grant local governments the authority to establish local subdivision standards, including street widths, the Uniform Fire Code, which was adopted by the State Fire Marshal, establishes a uniform street width across the United States. Realizing that such unwavering contingency would not be suitable to all environmental contexts, Oregon ruled that the width of streets established by local governments shall "supersede and prevail over any specifications and standards for roads and streets set forth in a uniform fire code."[29]

Another example of applying standards and codes at the local level of impact are transect-based codes. This approach to planning draws on the analytic concept of defining a cross-section of land that categorizes the natural and built environment as a continuum gradient along a transect. Introduced by the new-urbanism movement, codes associated with the transect focus on the conditions that maintain character and diversity within the typology of each distinct place. For example, under the transect-based codes streetscape at the most urban end of the transect may have a single species of tree in regularly spaced intervals, while at the more rural end, a gravel path may replace the sidewalk and the trees are arranged in clusters of multiple species. This approach contrasts with conventional codes, which emphasize simplification and use separation, but are devoid of generative forms associated with the character of the place.[30] Duany Plater-Zyberk & Company has codified this concept into a SmartCode©, which can be licensed and customized by various municipalities. To date, the most prominent use of transect zoning has been in Nashville-Davidson County, Tennessee, which incorporated a transect-based code into a 1999 revision of its zoning ordinance.

Consolidate and "Electrify" the Regulatory Approval Process   The red tape and bureaucratic procedures associated with development approval at the local level are due not only to the regulations themselves, but also to the multiple agencies and committees involved in the process. For example, Concord, Massachusetts—a small town of about 17,000 people—has at least 12 different commissions and committees involved in their approval process. To eliminate delays and jurisdictional conflicts, localities should consider consolidating their process and giving the final approval to an overseeing committee that may be composed of subcommittee representatives.

Realistically of course, while reducing a politicized approval track into the hands of fewer parties may be difficult to accomplish because of the current conflicted attitudes about development, streamlining the process through technological innovation can be more easily achieved. As Internet use spreads, and becomes more available, there is a growing expectation for the ability to conduct affairs from home or office with greater immediacy. From automatic approval of plans, to equipping inspectors with portable devices for recording and inspecting on-site, electronic permitting systems can provide better and timelier information to decision makers and experts alike. The possibility of electronic plan review is particularly encouraging for its potential to automatically analyze a plan, and compare it with codes and standards requirements. Alternatively, such systems can allow the plan reviewer to enter various descriptors and benchmarks, and let the software call up the applicable requirements that need to be considered. A recent HUD report, which strongly supports such systems in an effort to reduce regulatory barriers to housing, indicates that in jurisdictions that have implemented such systems, it has helped reduce turnaround time, in some cases by as much as 80 percent.[31]

Develop and Endorse a Local Vision Plan   The requirement that zoning and land-use decisions be made in accordance with a comprehensive plan remains one of the most ill-defined requirements of the Standard Zoning Enabling Act (SZEA) of 1922. The importance of such a plan in guiding development, however, has recently taken center stage.[32]

The renewed emphasis on urban growth and the debates about smart growth have shown that master plans are some of the most useful planning tools available to local governments.[33] The visions, goals, and aspirations of a community start with the translation of these concepts into a comprehensive vision plan. This plan becomes a leading instrument of place planning with supermajority approval by the community. The plan, endorsed and revised according to periodic needs, becomes the foundation for physical planning. Without such a vision plan, physical planning is reduced to an activity of implementing piecemeal, fragmented zoning regulations and subdivision deals that result in stereotypical American disjointed built form.

In contrast, out of a vision plan come precise plans for locales and districts. These plans describe in general the physical characteristics of an area and its compatible relationship with built form in adjoining districts, and then set standards for future development. These standards may include performance standards as well as specific prescriptions for streets, open space, structures, parking, and landscaping. A precise plan can also have a unique administrative section that specifies what levels of approval are needed for different permits, thus allowing for flexibility in the approval process.[34] Finally, communities should not overlook the importance of requiring specific site plans that truly address the local conditions of each site and respond to its overall context.

Envision Place Making   Finally, as we have seen in chapter 8, the public's difficulty in visualizing the physical ramifications of land-use and subdivision regulations is a barrier that must be overcome on the road to better local design and planning. Putting into use powerful yet readily available computational tools to introduce communities to the variety of choices available will help them visualize the potential effects that these choices produce, and will ultimately diversify the spatial paradigm of development.

Promising new venues can be seen in the application and adaptations of new technologies that are web based and do not require high-level computing skills or capabilities. The Visual Interactive Code (VIC)™, for example, is a computer-based Internet system that enables local governments to convert land-use regulations and planning data into a single visually based format using photographs, drawings, and maps. By utilizing an easy and engaging graphic interface (pictures and data that correlate to one another and are interchangeable), different effects of regulations can be shown. With a click of a mouse, end-users can view the configurations and layouts of various developments, density measurements, street widths and setbacks, as well as other related features.[35]

Afterword: Places Nonstandard

They always say time changes things, but you actually have to change
them yourself.

—Andy Warhol

In the last days of 2002, Habersham County, on the outskirts of metropolitan
Atlanta, abolished all its land-use regulations, fired all its building inspectors,
and eliminated its planning commission. "We're going to see if people truly
need to be regulated," said Commissioner Jerry Tanksley, who made the mo-
tion, to the *Atlanta Journal-Constitution*.[1]

Tanksley wanted to protest what he felt were burdensome development
restrictions placed on the county by federal and state agencies. The commission's action was the latest in a series of growth-related battles that have dominated local politics for years in Habersham County. No longer an isolated
region of mountain scenery and aging small towns, Habersham now has a four-
lane, limited-access highway connection to metro Atlanta. Opponents of the
increased regulations that have come with the growth have decried zoning and
planning as attempts to take away landowners' basic property rights.

The regulatory rollback followed years of battles between small land-
holders and outside regulators, dating back to the Georgia Planning Act of
1989 and, on a larger scale, to the days of the nineteenth-century mountain
men. One contentious issue was Atlanta's efforts to require a minimum buffer
zone between streams and new construction projects. The problem was not the
rule's immediate impact: the law did not apply to farmers, who still constituted
a substantial portion of the Habersham population. The problem was that outsiders were telling the county what to do, angering a citizenry that prized both
local self-government and individual property rights.

Surprisingly to some, the Home Builders Association of Habersham filed a lawsuit almost immediately after the abolition of the regulatory system, arguing that the county commission did not have the power to eliminate zoning laws. A Circuit judge agreed and promptly ordered the county to enforce the abandoned rules.

As in Habersham County, urban planning in the United States finds itself facing many challenges: from global economic pressures to the rise of locally based social and ethnic movements, to the ever-increasing demands for new housing and development. The profession even suffers from a sense of insecurity and self-paralyzing pessimism. To address and meet new challenges, we must allow for the fresh air of self-determination, for a clear vision of the destination toward which we should be heading, and for a flexible path to get us there. Local empowerment, the adoption of place-based guiding principles, and the renewed interest in urban form and design are already molding our future direction. Beyond that, versatility will be the key to reforming our regulatory paradigm.

Designers and planners and those who wish to shape the built environment cannot easily escape their obligation to create and maintain places where rules do not oust the physical qualities that attract and anchor people. We must adhere to notions of "goodness" as a test for regulations and forming standards and at the same time ensure that adequate flexibility is provided for ever-changing markets, lifestyle demands, and personal values. The professionals among us must make certain that standards do not result in mediocrity in urban form or in the public realm. They must realize that good urban design is not a result of standards and regulations derived from mathematical formulas, but rather from experience gained from use.

Regulations and standards will continue to exert influence and shape the built form of places. As in the case of common-interest communities, private land-use controls are regularly being utilized and agreed to by the local residents in such developments. Acceptance of these private controls indicates recognition of the need for norms to enhance and establish quality. These self-imposed, locally generated sets of standards and codes reflect a societal adaptation in response to the failures associated with the ubiquitous nature of conventional regulations.

The standards we have to work with should only be used as a baseline and not as a device to prevent excellence from being created. Above all, even at the risk of being perceived as doctrinaire, we must take a formal stand against the adoption of rules that perpetuate mediocre development outcomes. There must be a willingness to test standards in relation to their impact on the form of communities and place making.

A key goal of any planning effort is to create places that are sustainable and well-designed, places where optimal quality and efficiency in the provision and arrangement of urban amenities and services are attained. However, place making today stands in poor relationship to civic processes and urban design. It has been crowded into a bureaucratic task of rule writing, standards formation, and code enforcement. Such roles rob the urban planning profession of its central goal: to foster democratic civic processes and outcomes whereby communities retain their local character, make the most of the existing conditions of the built and natural environment, and create developments that are sensitive and sensible to their immediate surroundings.

In the last decade, we have created a genetic bank that promotes cloning rather than mutation. The process of producing multiple sets of standards, all practically identical in terms of a single ancestor, and applying them with disregard to place and locale, has more often than not created ubiquitous, unsympathetic spaces.

Standards must be place-base-tested. Standards must be allowed to evolve. And to evolve standards we must allow for experimentation and discretion.

Alexis de Tocqueville observed on his tour of the United States in the 1830s that "the great privilege enjoyed by the Americans is not only to be more enlightened than other nations but also to have the chance to make mistakes that can be retrieved."[2]

Taking chances, allowing experimentation, and letting experts and regulators use their judgment are practices that must find their way back into the planning process. The hope may rest with the upcoming generation, tuned and responsive to the natural environment, with an understanding of the interaction between socioeconomic issues and spatial design.

Appendix A

The Code of the City—Timeline

2000 BC	The religious writings of the Vedas specify the city law of the Indus Valley civilization.
2000 to 1000 BC	The towns of Kahun and Tel El-Amarna in Egypt are laid out in a formal pattern.
1700 BC	Hammurabi's Code is issued.
1400 BC	Clay tablets from Sumerian culture show records of land measurements and plans for agricultural and built areas.
350 BC	The Chinese Code of Li k'vei is formulated.
350 BC	Greek cities pass bylaws to secure the public order of markets and streets.
221 BC	A unified land-measurement system is developed in China by the first emperor, Shin Huang-Ti.
200 BC	Regulations for the construction of cities and buildings are mentioned in the *Zhouli,* one of the Confucian Classics, from the Zhou Dynasty.
100 BC to 220 AD	The Chinese walled ward system (*li*) is implemented as a dimensional mechanism for town planning.
104 to 43 BC	The charter of the municipality of Tarentum (present-day Italy) deals with the unlawful destruction of buildings, typical of design guidelines of the time.
40 BC	Architect Marcus Vitruvius Pollio writes the handbook *De architecture libri decem* (Ten Books on Architecture), covering both good architecture and design standards.

31 BC to 14 AD	The Roman Emperor Augustus Caesar limits the height of buildings to avoid dark, narrow passages.
64	The Roman Emperor Nero limits the height of dwellings to 70 feet.
529	Justinian's Code is issued.
531	Julian of Ascalon writes a treatise covering land use, views, house construction, drainage, and planting.
1100 to 1200	Islamic cities are regulated by Islamic law emphasizing social behavior.
1262	Siena (as well as other European cities) enacts statutes to control building in a defensive zone adjacent to the city's defensive wall.
1542	The New Laws of the Indies prescribe the characteristics of new towns.
1548	A Paris law is enacted to contain development within the city, forbidding the construction of new homes in the *faubourgs* (outlined areas).
1585	Sixtus V transforms Rome.
1606	Edmond Gunter develops an accurate and easy-to-use land-measurement system.
1666	The Great Fire devastates London.
1667	The London Building Act is passed imposing restrictions on building height.
1699	The Plan of Williamsburg, Virginia, is adopted; it had a rectangular and formal layout with a market square, open spaces, and street vistas.
1721	The U.S. Act of "surveyors and regulators" is passed to establish streets and building lines in Philadelphia.
1766	Governor Luis Antonio de Souza of Portugal stipulates uniformity and order in the laying out of new towns and cities.
1774	The Building Act to regulate development of buildings and manage waste takes effect in London.

1785	The U.S. Land Ordinance for the Western Territory is passed.
1787	The U.S. Congress passes an ordinance to establish the boundaries of the states of Illinois, Indiana, and Ohio, which surveyed the land in 640-acre sections.
1794	*A General Plan for Laying Out Towns and Townships on the New-Acquired Lands in the East Indies, America, or Elsewhere* by Granville Sharp appears.
1811	The peninsula of Manhattan is laid out in grid form by the Commissioners, Gouverneur Morris, Simeon de Witt, and John Rutherford.
1825	The Building Act to regulate development of buildings and manage waste takes effect in Liverpool, England.
1835	In England, a 6-foot height for sewers is stipulated to allow people to walk upright through the sewers to clean them.
1842	The *Report on The Sanitary Conditions of The Labouring Population of Great Britain* is produced.
1844	The *First Report of the Commissioners of the State of Large Towns and Population Districts* appears in England.
1844	The Building Act establishes town-planning principles in England.
1848	The Public Health Act is passed and the General Board of Health is established in England.
1855	The Paris boulevards are created by Napoléon III and Georges-Eugene Haussmann.
1855	The first "model tenement" is built in New York City.
1860	The first professional engineering education is offered at land-grant institutions in the United States.
1865	Italian regulations (*piano regolatore* and *piano di ampliamento*) are introduced, required for the design of existing and new areas in cities with a population of 10,000 or more.
1867	The first tenement-house law is passed in New York City.

1867   San Francisco enacts the first ordinance in the United States that could be classified as zoning. This ordinance merely addresses the location of obnoxious uses.

1867   The National Society of Civil Engineers is established in the United States.

1868   The Institution of Surveyors is formed in London.

1875   The Public Health Act ordinance is passed, The "Bye Law" Street, England.

1875   Prussian law is enacted concerning the laying out of and alteration of streets and squares in cities and country places.

1885   The City of Modesto, California, bans steam laundries in residential areas.

1887   Tenement-house law is enacted in New York City requiring that running water be placed on each floor; it also stipulates that only 65 percent of a lot is permitted to be covered by buildings.

1887   The Model By-Laws are passed in England.

1889   The metropolitan Sewerage Commission is created in the Boston area, which covered eighteen towns.

1890   The *Handbuch des Städtebaue,* written by Josef Stübben, is published.

1891   The Boston Board of Survey Act is passed.

1893   The World's Columbian Exposition in Chicago showcases utopian visions, becoming a source of the city beautiful movement and of the urban planning profession.

1895   The Metropolitan Water Act, applying to the Boston area, is passed.

1899   The U.S. Congress sets building-height restrictions in Washington, D.C. (60 feet for nonfireproof residential buildings, 90 feet maximum for all other buildings on residential streets, a maximum of 130 feet for all buildings on all other streets).

1901   The New York State Tenement House Law is enacted, becoming the legislative basis for the revision of city codes outlawing certain tenements.

1904   Boston building-height restrictions are introduced (for wooden buildings, a maximum of 45 feet; for other buildings: in the business districts, 125 feet; in other parts of the city, a maximum of 80 feet).

1909   The first city planning conference is held in Washington, D.C.

1909   Los Angeles adopts an ordinance creating seven industrial districts and zoning the rest as residential districts; it becomes the first municipality to apply zoning to undeveloped land.

1909   The Housing, Town Planning, (Etc.) act is passed in England.

1909   The first national U.S. conference on city planning is held in Washington, D.C.

1910   The National Housing Association is organized in the United States.

1911   *The Principles of Scientific Management,* by Frederick Winslow Taylor, appears.

1913   New Jersey becomes the first state to institute the mandatory referral of subdivision plats. This is the beginning of modern subdivision controls.

1914   Leonard Metcalf and Harrison P. Eddy's textbook *American Sewerage Practice* is published.

1916   New York City enacts the first modern, "comprehensive" zoning ordinance—the first zoning ordinance to contain land-use, density, and building-bulk controls.

1916   The U.S. Federal Aid Roads laws are established to support the development of transcontinental highways.

1917   The City Planning Institute is formed in Kansas City during the 9th National Conference on City Planning.

1918   The Massachusetts State Constitution has a new amendment enabling the legislature to limit the use and construction of specific towns and areas.

1919   The U.S. Bureau of Industrial Housing and Transportation publishes its *Report of the United States Housing Corporation,* Volume 2: *Houses, Site-planning, Utilities.*

1920   The U.S. Federal Power Commission Act is set up to collect and record data on regional water resources and is also empowered to issue licenses for development in various areas.

1920   The Interstate Commerce Commission prepares a tentative plan to consolidate the railways in the United States.

1921   The Federal Highway Act is passed to fund a system of connected highways in the United States.

1922   The U.S. Department of Commerce publishes the Standard State Zoning Enabling Act (SZEA). This is a model law that when passed by a particular state would enable that state's municipalities to enact zoning ordinances.

1922   The Interim Development Orders to control development are introduced in England.

1926   In *Village of Euclid, Ohio v. Ambler Realty Company,* the U.S. Supreme Court confirms the constitutionality of zoning.

1928   The U.S. Department of Commerce publishes the Standard City Planning Enabling Act (SCPEA), a model law that when passed by a particular state would enable that state's municipalities to adopt it.

1929   *The Neighborhood Unit* by Clarence Perry is published.

1930   The U.S. National Institute of Transportation Engineers is formed.

1931   President Herbert Hoover's Conference on Home Building and Home Ownership takes place.

1934   *The Design of Residential Areas* by Thomas Adams appears.

1934   The U.S. Federal Housing Administration (FHA) is established.

1935   The FHA's *Subdivision Development* (National Housing Act) is introduced.

1935   Standards for the Insurance of Mortgages on Properties Located in Undeveloped Subdivisions—Title II of the National Housing Act—become law.

1935   The FHA's *Property Standards and Subdivision Development Standards* is published.

1936   The FHA's *Planning Neighborhoods for Small Houses* appears.

1936   The *Model Subdivision Regulations,* Advisory Committee on City Planning and Zoning, United States, is produced.

1936   *Model Laws for Planning Cities, Counties and States, Including Zoning, Subdivision Regulations, and Protection of Official Map* by Edward Bassett appears.

1938   The FHA's *Subdivision Standards* are introduced.

1938   *Basic Principles of Healthful Housing,* Committee on Hygiene and Housing Public Health Association, United States, are formulated.

1939   *Standards for Modern Housing,* Public Health Association, are introduced.

1939   *Land Subdivision,* American Society of Civil Engineers, is established.

1942   *Traffic Engineering Handbook,* Institute of Traffic Engineers (ITE), is produced.

1942   *The Subdivision of Land: A Guide for Municipal Officials,* American Society of Planning Officials, is issued.

1942   *Standards for War Housing,* U.S. National Housing Agency, are developed.

1947   *A Checklist for the Review of Local Subdivision Controls,* U.S. National Housing Agency, is produced.

1947   The Town and Country Planning Act, England, is passed.

1948   *Planning the Neighborhood,* American Public Health Association, is issued.

1949   Housing Act, United States—the use of eminent domain is introduced.

1952   The *Manual* of the U.S. Housing and Home Finance Agency calls for more widespread subdivision controls.

1957   *Urban Land Use Planning,* by Stuart Chapin, is published.

1960   *Recommended Practice for Subdivision Streets,* ITE, appears.

1962   Fairfax County, Virginia: State Board of Supervisors establishes Virginia's first residential planned community zone, the precursor of Planned Unit Development (PUD).

1962   In France, "Loi Malraux" is the first of the historic preservation laws to protect historic cores from urban renewal. It is followed by England's Civic Amenities Act of 1967 and the U.S. National Historic Preservation Act of 1966.

1970   The National Environmental Policy Act (NEPA) is signed in the United States.

1972   The U.S. Federal Water Pollution Control Act amendment is passed to subsidize construction of local treatment works.

1974   The U.S. Environmental Protection Agency (EPA) identifies noise levels affecting health and welfare.

1976   The *Model Land Development Code,* American Law Institute, is formulated.

1977   The U.S. Clean Water Act is passed.

1980   *Performance Zoning,* by Lane Kending, is published.

1986   The U.S. National Urban Runoff Program is established.

1987   The U.S. Water Quality Act is passed.

1991   *Not in My Back Yard: Removing Barriers to Affordable Housing*—a report by the U.S. Advisory Commission on Regulatory Barriers to Affordable Housing—is issued.

1991   The Planning and Compensation Act, England, is passed.

1994   The International Code Council (ICC) is established.

1996   *Skinny Streets, Better Streets for Livable Communities,* Livable Oregon, and the Transportation and Growth Management Program appear.

1999   *Traditional Neighborhood Development Street Design Guidelines,* Institute of Transportation Engineers, are developed.

2000	ICC's International Residential Code (IRC), an International Private Sewage Disposal Code (IPSDC), an International Property Maintenance Code (IPMC), an International Zoning Code (IZC), and even an International Urban-Wildland Interface Code (IUWIC) are formulated.
2002	The *Growing Smart Legislative Guidebook,* American Planning Association (APA), is published.
2002	The U.S. Department of Housing and Urban Development Regulatory Barriers Clearinghouse is established.
2002	*Context Sensitive Design,* U.S. Federal Highway Administration (FHWA) Memorandum, is issued.
2002	Form Based Codes are approved in Columbia Pike, Arlington, Virginia.
2003	A report titled *The Practice of Low Impact Development* is published by the U.S. Department of Housing and Urban Development.
2005	*SmartCode User's Manual,* published by Duany Plater-Zyberk & Company

Appendix B

The Code of the City—Matrix

Type of Code	Characteristics	Examples	Advantages	Disadvantages
Conventional zones and districts (Euclidean)	Includes "districts," "uses," and "dimensional and density standards." "Proscriptive": prohibits development not consistent with the code. Generally text-based with mapped districts.	Base districts Use classifications Dimensional standards: setbacks, height, lot size, density, floor-area ratio.	Fairly easy for staff to implement and for the public to interpret, if well organized. Familiar to professionals, staff, public officials, and the public. Flexibility for varied design within parameters of use and dimensional standards. Results are predictable.	Lack of flexibility to address different site characteristics and surroundings. Often disregards existing development patterns, creating numerous nonconformities. Does not prescribe qualitative development outcome, allowing for uncertainty as to product design. Generally text-based, difficult for the public to interpret the physical consequences, particularly if not well organized.
Planned development	Allows flexibility from standard rules to permit mixed uses, creative design, and/or public benefits. Highly discretionary (negotiation) with findings usually required.	"Planned development" zones. "Planned unit development" allowances and districts. "Planned community" zones.	Flexibility to allow creative design, mixed uses, to achieve preferred site development and public benefits. Ability to forecast and see final plan and design solutions over time.	Highly discretionary process leads to high degree of uncertainty. Negotiations may result in perceptions of public "giveaways" to or unreasonable "exactions" from developer.

Type of Code	Characteristics	Examples	Advantages	Disadvantages
Performance standards	Regulates development "impacts" such as nuisance factors, impervious surface, landscape surface area, trip generation, etc.	Nuisance (odor, noise, vibration, glare, toxics, etc.) standards in industrial or commercial zones. Performance criteria (floor area, impervious surface, trip generation, etc.) to compare development alternatives.	All parties involved in process and solution. Flexibility to vary uses, density, and intensity of development and to address impacts instead.	Time-consuming; may require central management and covenants. Impact approach may not address site-specific conditions or constraints. Difficult to implement—requires complex calculations.
Incentive-based codes and guidelines	Flexibility to achieve objectives through "incentives" such as density or floor-area bonuses in exchange for provision of selected uses and public amenities.	Exemption from FAR for child care or cultural uses. FAR bonuses for preserving historic structures. Density bonuses for affordable housing.	Optional to developer. Relies on a carrot rather than stick approach. May provide public amenities with "win-win" approach	Incentives may not be used, and amenities not provided. Bonus and benefits offered may be perceived as excessive.

| Form-based codes or design-oriented codes and districts | Graphic-based and design approach to outlining regulations, including design "typologies" for homes, shop fronts, commercial areas, public spaces, streetscapes, etc.

"Prescriptive": outlines what is expected of development, especially design.

Uses and dimensional standards downplayed.

"Regulating plan" to outline design typologies. | Traditional neighborhood development (TND) zone.

Urban village zone.

Neighborhood marketplace zone.

Transit-oriented development (TOD) zone. | Graphics and illustrations more readily understood by public, public officials, developers, and professionals.

Preferred or required development approaches conceptually rendered or illustrated.

Integrates principles of mixed-use and pedestrian orientation.

Useful for developing areas and redevelopment or infill sites. | Not readily applicable to built-out urban or suburban areas.

Requires significant, upfront effort to develop regulating plan and design specifics.

Perception of initial market resistance.

May not provide enough design flexibility to applicants. |

Source: Partly based on "Types of Zoning Codes and Formats." Discussion Paper. Palo Alto, CA: City of Palo Alto Department of Planning and Community Environment, 2001.

Notes

Chapter 1

1. Quoted in "Designing Black Rock City," the Burning Man website (http://www.burningman.com/whatisburningman/about_burningman/brc_growth.html, December 2003).

2. Thomas Hobbes, *Leviathan* (Baltimore, MD: Penguin Books, 1968), 99b.

3. Binode Dutt, *Town Planning in Ancient India* (Calcutta: Thacker, Spink & Co., 1925), 58.

4. Dutt, *Town Planning in Ancient India,* 248.

5. Dutt, *Town Planning in Ancient India,* 248–257.

6. Dutt, *Town Planning in Ancient India,* 143.

7. Nancy Steinhardt, *Chinese Imperial City Planning* (Honolulu: University of Hawaii Press, 1990), 33.

8. Steinhardt, *Chinese Imperial City Planning,* 96.

9. http://fischer.jinkan.kyoto-u.ac.jp/soramitsu/addressing.html.

10. Francis Haverfield, *Ancient Town-Planning* (Oxford: Clarendon Press, 1913), 37.

11. Plato, *Laws,* trans. R. G. Bury (Cambridge, MA: Harvard University Press, 1926), book VI, 778 B, D; 779 B, C, D.

12. Francis Haverfield, *Ancient Town-Planning* (Oxford: Clarendon Press, 1913), 54.

13. August Mau, *Pompeii, Its Life and Art* (New York: Macmillan, 1899).

14. Thomas Adams, *Outline of Town and City Planning* (New York: Russell Sage Foundation, 1936), 60.

15. Haverfield, *Ancient Town-Planning,* 73.

16. Adams, *Outline of Town and City Planning,* 53.

17. *Roman Laws and Charters, and Three Spanish Charters and Other Documents,* translated with introduction and notes by Ernest George Hardy. (New York: Arno Press, 1975), 101.

18. For an excellent discussion of Julian's treatise, see Besim S. Hakim, "Julian of Ascalon's Treatise of Construction and Design Rules from Sixth-Century Palestine," *Journal of the Society of Architectural Historians* 60/1 (March 2001), 4–25.

19. Saleh Al-Hathloul, *Tradition, Continuity, and Change in the Physical Environment: The Arab-Muslim City,* doctoral dissertation, MIT, 1981, 107.

20. Al-Hathloul, *Tradition, Continuity, and Change in the Physical Environment,* 137.

21. Al-Hathloul, *Tradition, Continuity, and Change in the Physical Environment,* 84.

22. Al-Hathloul, *Tradition, Continuity, and Change in the Physical Environment,* 92.

23. Besim Hakim, *Arabic-Islamic Cities: Building and Planning Principles* (London: Kegan Paul International, 1986); Besim Hakim, "The 'Urf' and Its Role in Diversifying the Architecture of Traditional Islamic Cities," *Journal of Architectural and Planning Research* 11/2 (1994), 108–127.

24. Robert Lopez, "The Crossroads within the Wall," in Oscar Handlin and John Burchard, eds., *The Historian and the City* (Cambridge: MIT Press and Harvard University Press, 1966), 30–35.

25. Howard Saalman, *Medieval Cities* (New York: George Braziller, 1968), 23.

26. Saalman, *Medieval Cities,* 30–31.

27. Spiro Kostof, *The City Assembled: The Elements of Urban Form through History* (Boston: Bulfinch Press, 1992), 31.

28. Thomas Tout, *Medieval Town Planning: A Lecture* (Manchester: Manchester University Press, 1934), 12–31.

29. Anthony Garvan, "Proprietary Philadelphia as an Artifact," in Oscar Handlin and John E. Burchard, eds., *The Historian and the City,* 177–201 (Cambridge, MA: MIT Press, 1966).

30. Doxiadis, Riyadh Master Plan, A-19; quoted in Al-Hathloul, *Tradition, Continuity, and Change in the Physical Environment,* 174.

31. Directive No. 381/5, dated 16/11/1395/1975; quoted in Al-Hathloul, *Tradition, Continuity, and Change in the Physical Environment,* 187.

Chapter 2

1. Spiro Kostof, *The City Shaped: Urban Pattern and Meaning through History* (London: Bulfinch, 1991), 256.

2. Kostof, *The City Shaped,* 258.

3. Kostof, *The City Shaped,* 256.

4. Quoted in Kostof, *The City Shaped,* 256.

5. Kostof, *The City Shaped,* 258.

6. Francis Thompson, *Chartered Surveyors: The Growth of a Profession* (London: Routledge, 1968), 64.

7. Clara Greed, *Surveying Sisters: Women in a Traditional Male Profession* (London: Routledge, 1991), 47.

8. Thompson, *Chartered Surveyors,* 140.

9. Quoted in Clifford Knowles and Peter Pitt, *The History of Building Regulation in London 1189–1972* (London: Architectural Press, 1972), 61.

10. Patrick Abercrombie, *Town and Country Planning* (London: Oxford University Press, 1933), 78.

11. Thompson, *Chartered Surveyors,* 128.

12. Thompson, *Chartered Surveyors,* 180.

13. Robert Home, *Of Planting and Planning: The Making of British Colonial Cities* (London: Spon Press, 1997), 40.

14. Home, *Of Planting and Planning,* 40.

15. Lewis Mumford, *The City in History: Its Origins, Its Transformations, and Its Prospects* (New York: Harcourt, Brace & World, 1961), 421.

16. U.S. Department of Agriculture, *Land Use and Its Pattern in the United States* (Washington, DC: USDA); quoted in Norman Thrower, *Original Survey and Land Subdivision: A Comparative Study of the Form and Effect of Contrasting Cadastral Surveys* (Chicago: Rand McNally, 1966), 6.

17. See the NASA website "Earth from Space and Human Interaction" at http://earth.jsc .nasa.gov/sseop/efs/mm.htm#.

18. Norman Thrower, *Original Survey and Land Subdivision: A Comparative Study of the Form and Effect of Contrasting Cadastral Surveys* (Chicago: Rand McNally, 1966), 96.

19. Andro Linklater, *Measuring America: How an Untamed Wilderness Shaped the United States and Fulfilled the Promise of Democracy* (New York: Walker, 2002, 146).

20. Darrel Miller, *Life on the Central Branch.* Railroadtown.com. Available at http://www .downsnews.com/railroadtown/.

21. Francis Haverfield, *Ancient Town-Planning* (Oxford: Clarendon Press, 1913), 42.

22. Polybius, *Histories* (200–118 BCE), book 6, 28–31. Online text of the Cultural Heritage Language Technologies. http://www.chlt.org/cache/perscoll_Greco-Roman.html.

23. John Rae, *The Road and the Car in American Life* (Cambridge, MA: MIT Press, 1971), 38.

24. Rae, *The Road and the Car in American Life,* 74.

25. Daniel Calhoun, *The American Civil Engineer: Origins and Conflict* (Cambridge, MA: Technology Press, MIT, 1960).

26. Theodore Matson and Wilbur Smith, *Traffic Engineering* (New York: McGraw-Hill, 1955), 3.

27. Institute of Transportation Engineers, *Recommended Practices for Subdivision Streets* (Washington, DC: ITE, 1967), 6.

28. Organization for Economic Co-Operation and Development, *Geometric Road Design Standards* (Paris: OECD, 1977), 196.

Chapter 3

1. Carol Cristensen, *The American City: Concepts and Assumptions,* doctoral dissertation, University of Minnesota, 1977.

2. Antrim Haldeman, "The Control of Municipal Development by the 'Zone System' and Its Application in the United States," *Proceedings of the Fourth National Conference on City Planning, Boston, Massachusetts, May 27–29, 1912,* 173–188 (Boston: National Conference on City Planning, 1912).

3. Frederic C. Howe, "The City as a Socializing Agency: The Physical Basis of the City: The City Plan," *American Journal of Sociology* 17 (March 1912), 590–601 (quote on 590–591).

4. Quoted in Frederick Winslow Taylor, *Scientific Management, Comprising Shop Management: The Principles of Scientific Management and Testimony before the Special House Committee.* New York: Harper & Row, 1911. Introduction (online version, at http://www.marxists.org/reference/subject/economics/taylor/principles/index.htm September 2003).

5. Taylor, *Scientific Management, Comprising Shop Management.*

6. E. P. Goodrich and George Ford, "Efficiency in City Planning," *The American City,* February 1913, 139–142 (quote on 139).

7. Goodrich and Ford, "Efficiency in City Planning," 142–143.

8. George Ford, *Engineering Record* 67 (May 17, 1913), 551–552.

9. Ford, *Engineering Record* 67, 551.

10. See Mel Scott, *American City Planning Since 1890* (Berkeley: University of California Press, 1971), 166–167.

11. See Frederic Howe, "The Municipal Real Estate Policies of German Cities," *Proceedings of the Third National Conference on City Planning, Philadelphia, Pennsylvania, May 15–17, 1911*, 14–26 (Cambridge: Cambridge University Press, 1911). The quotation is from his more refined article; see Howe, "The City as a Socializing Agency" (quote on 593).

12. See Josef Stübben, "Practical and Aesthetic Principles for the Laying Out of Cities," first presented at a meeting of the Deutschen Vereins für öffentliche Gesundheitspflege, held in Freiburg, Germany, in September 1885. A translation by W. H. Searles was issued as an Advance Copy of a portion of the *Transactions* of the American Society of Civil Engineers of 1893. Its publication was part of the meeting in Chicago of the International Engineering Congress of the Columbian Exposition, 1893 (available at http://www.library.cornell.edu/Reps/DOCS/stubb_85.htm).

13. Stübben, "Practical and Aesthetic Principles for the Laying Out of Cities."

14. See Michael Southworth and Eran Ben-Joseph, *Streets and the Shaping of Towns and Cities* (Washington, DC: Island Press, 2003).

15. Howe, "The City as a Socializing Agency," 595.

16. Robert Whitten and Thomas Adams, *Neighborhoods of Small Homes* (Cambridge, MA: Harvard University Press, 1931), 87.

17. John Gries and James Ford, eds., *Planning Residential Districts: Report on the President's Conference on Home Building and Home Ownership* (Washington, DC: National Capitol Press, 1932), 1–2.

18. Gries and Ford, *Planning Residential Districts,* 124.

19. Harold Lautner, *Subdivision Regulation: An Analysis of Land Subdivision Control Practices* (Chicago: Public Administration Service, 1941), 1.

20. Massachusetts Subdivision Control Law, MGL, chap. 41, section 81K.

21. Ebenezer Howard, introduction to *Garden Cities of To-morrow* (London: Faber and Faber, 1945).

22. Harold Stubblefield and Patrick Keane, *Adult Education in the American Experience: From the Colonial Period to the Present* (San Francisco: Jossey-Bass, 1994), 173–174.

23. Mel Scott, *American City Planning Since 1890* (Berkeley: University of California Press, 1969), 73.

24. Alfred Yeomans, *City Residential Land Development: Competitive Plans for Subdividing a Typical Quarter Section of Land in the Outskirts of Chicago* (Chicago: University of Chicago Press, 1916), 1.

25. Yeomans, *City Residential Land Development,* 39.

26. Marlene Shore, *The Science of Social Redemption: McGill, the Chicago School, and the Origins of Social Research in Canada* (Toronto: University of Toronto Press, 1987), 111.

27. James Dahir, *The Neighborhood Unit Plan, Its Spread and Acceptance* (New York: Russell Sage Foundation, 1947).

28. Clarence Perry, "The Neighborhood Unit: A Scheme of Arrangement for the Family Community," in *Regional Plan of New York and Its Environs* VII (New York: Regional Plan Association, 1929), 34.

29. Perry, "The Neighborhood Unit," 34–45.

30. Clarence Perry, *Housing for the Machine Age* (New York: Russell Sage Foundation, 1939), 26.

31. U.S. Federal Housing Administration, *Planning Neighborhoods for Small Houses,* Technical Bulletin No. 5 (Washington, DC: FHA, July 1, 1936), 28.

32. Tridib Banerjee and William Baer, *Beyond the Neighborhood Unit* (New York: Plenum Press, 1984).

33. See "APHA's History," a series of six articles reflecting on APHA's history as the association marked its 125th anniversary. The articles appeared in issues 4, 5, 6, 8, 9, and 10 in *The Nation's Health* 27 (1997).

34. "APHA's History" (see note 33 for the full reference).

35. American Public Health Association, Committee on Hygiene of Housing, *Planning the Neighborhood* (New York: APHA, 1948), 1.

36. American Public Health Association, Committee on Hygiene of Housing, *Planning the Neighborhood,* 12–16.

37. U.S. Federal Housing Administration, *The FHA Story in Summary, 1934–1959* (Washington, DC: FHA, 1959), 12–16.

38. Marc Weiss, *The Rise of the Community Builder* (New York: Columbia University Press, 1987), 152.

39. U.S. Federal Housing Administration, *Subdivision Development,* Circular No. 5 (Washington, DC: FHA, January 10, 1935), 3.

40. U.S. Federal Housing Administration, *Planning Neighborhoods for Small Houses,* Technical Bulletin No. 5 (Washington, DC: FHA, July 1, 1936), 9.

41. U.S. Federal Housing Administration, *Planning Neighborhoods for Small Houses,* 12.

42. U.S. Federal Housing Administration, *Successful Subdivisions,* Land Planning Bulletin No. 1 (Washington, DC: FHA, 1940), 3.

43. Marc Weiss, *The Rise of the Community Builder* (New York: Columbia University Press, 1987), 153.

44. International City Managers Association, *Local Planning Administration,* Municipal Management Series (Chicago: International City Managers Association, 1941), 256.

Chapter 4

1. Jack Newville, *New Engineering Concepts in Community Development* (Washington, DC: Urban Land Institute, 1967), 27.

2. Richard Tabors, Michael Shapiro, and Peter Rogers, *Land Use and the Pipe* (Lexington, MA: Lexington Books, 1976).

3. International City Managers Association, *Local Planning Administration,* Municipal Management Series (Chicago: ICMA, 1941), 146.

4. U.S. General Accounting Office, *U.S. Infrastructure Funding Trends and Federal Agencies' Investment Estimates* (Washington, DC: General Accounting Office, 2001).

5. In Sim Van der Ryn, *The Toilet Papers: Designs to Recycle Human Waste and Water: Dry Toilets, Greywater Systems and Urban Sewage* (Santa Barbara: Capra Press, North Atlantic Books, 1978), 1.

6. American Water Works Association, *Residential End Uses of Water 1999* (Denver: AWWA, 1999).

7. Janusz Niemczynowicz, "Water Management and Urban Development: A Call for Realistic Alternatives for the Future," *Impact of Science on Society,* UNESCO 166 (1992), 131–147.

8. "Ancient Hygiene," *The Sanitary Engineer and Construction Record,* January 1887, 179–180.

9. See Leonard Metcalf and Harrison Eddy, *American Sewerage Practice* (New York: McGraw-Hill, 1914), 1–2, 10–14; George Tchobanglous, *Wastewater Engineering: Collection and Pumping of Wastewater* (New York: McGraw-Hill, 1981), 2–3.

10. Metcalf and Eddy, American Sewerage Practice, 4.

11. Mary Gayman, "Glimpse into London's Early Sewers," *Cleaner Magazine*, COLE Publishing Inc., 1996. Retrieved from: http://www.swopnet.com/engr/index.html

12. Edwin Chadwick, *Report on the Sanitary Conditions of the Labouring Population of Great Britain*, ed. Michael Walter Flinn (Edinburgh: Edinburgh University Press, [1842] 1965), 16.

13. General Board of Health, *Information Collected with Reference to Works for the Removal of Soil Water or Drainage of Dwelling Houses and Public Edifices and for the Sewerage and Cleansing of the Sites of Towns* (London: Her Majesty's Stationery Office, 1852), 13.

14. Quoted in Sharon Beder, "Early Environmentalists and the Battle against Sewers in Sydney," *Journal of the Royal Australian Historical Society* 76/1 (1990), 27–44 (quote on 30).

15. Quoted in *Engineering Magazine* 1 (January 1869), 44.

16. Edward Philbrick, "Lecture I: Introductory," *American Sanitary Engineering* (New York: The Sanitary Engineer, 1881), 1–15 (quote on 10–11).

17. *Engineering Magazine* 1 (October 1869), 882.

18. Edward Philbrick, "Modern Sewer Construction and Sewage Disposal," *The Sanitary Engineer and Construction Record*, March 1887, 345–346.

19. Philbrick, "Modern Sewer Construction and Sewage Disposal," 346.

20. Mansfield Merriman, *Elements of Sanitary Engineering* (New York: Wiley, 1898), 149–150.

21. George E. Waring Jr., "Liernur's Pneumatic System of Sewerage," *Atlantic Monthly* 37/222 (April 1876), 482.

22. "The Shone Hydro-Pneumatic System of Sewerage," *Manufacturer and Builder*, May 1887, 104–105.

23. Merriman, *Elements of Sanitary Engineering*, 173.

24. Waring, "Liernur's Pneumatic System of Sewerage," 492.

25. Rudolph Hering, *Reports of an Examination Made in 1880 of Several Sewerage Works in Europe*, Annual Report of the National Board of Health (Washington, DC: U.S. National Board of Health, 1882), 136.

26. In Sharon Beder, "Pipelines and Paradigms: The Development of Sewerage Engineering," *Australian Civil Engineering Transactions* CE35/1 (March 1993), 79–85 (quote on 82).

27. Joel Tarr, J. McCurley, F. Michael, and T. Yosie, "Water and Wastes: A Retrospective Assessment of Wastewater Technology in the United States, 1800–1932," *Technology and Culture* 25/2 (April 1984), 226–263.

28. Quoted in Beder, "Pipelines and Paradigms," 82.

29. Willard Bascom, "The Disposal of Waste in the Ocean," *Scientific American* 231/2 (August 1974), 16–25 (quote on 24).

30. To understand the Polya process, Arthur suggested an example of an urn with infinite space containing one red ball and one white ball. If one were to randomly select one ball, there is a 50 percent probability that one would select either a red or a white ball. If one then proceeded to sample with replacement, but with each replacement added another ball of the same color as the ball replaced, one would be sampling with double replacement. Therefore, if one had selected a red ball in the first round, and after replacement, two red balls and one white ball were in the urn, the corresponding probabilities of selection are 66.6 and 33.3 percent. The Polya-Eggenberger model is derived by drawing balls of two different colors from an urn. As the balls are drawn, they are not only replaced, but new balls of the same color are added. In this way, numerous drawings of balls of one color greatly increase the probability of that color being drawn. (Description is based on Anthony Woodlief, "The Path-Dependent City," *Urban Affairs Review* 33/3 (January 1998), 405–437 (quote on 407).

31. W. Brian Arthur, "Competing Technologies, Increasing Returns, and Lock-In by Historical Events," *Economic Journal* 97 (1989), 642–665; W. Brian Arthur, "Positive Feedbacks in the Economy," *Scientific American* 262 (February 1990), 92–99.

32. Stan Liebowitz and Stephen Margolis, "Path Dependence, Lock-In and History," *Journal of Law, Economics and Organization* (April 1995).

33. W. Brian Arthur, *Increasing Returns and Path-Dependence in the Economy* (Ann Arbor: University of Michigan Press, 1994).

34. Robin Cowan, "Tortoises and Hares: Choice among Technologies of Unknown Merit," *Economic Journal* 101/407 (1991), 801–814.

35. Paul David, "Clio and the Economics of QWERTY," *American Economic Review* 75/2 (1985), 332–337.

36. Anthony Woodlief, "The Path-Dependent City," *Urban Affairs Review* 33/3 (January 1998), 405–437.

37. In Sharon Beder, "Technological Paradigms: The Case of Sewerage Engineering," *Technology Studies* 4/2 (1997), 168.

38. Joel Tarr and Gabriel Dupuy, eds., *Technology and the Rise of the Networked City in Europe and America* (Philadelphia: Temple University Press, 1988); Sharon Beder, *The New Engineer: Management and Professional Responsibility in a Changing World* (South Yarra, Australia: Macmillan Education Press, 1998).

39. Joel Tarr, J. McCurley, F. Michael, and T. Yosie, *Retrospective Assessment of Waste Water Technology in the United States: 1800–1972*, a report to the National Science Foundation

(San Francisco: San Francisco Press, 1977), 1; Martin Melosi, *The Sanitary City: Urban Infrastructure in America from Colonial Times to the Present* (Baltimore: Johns Hopkins University Press, 2000).

40. See chap. 2 of Richard Tabors, Michael Shapiro, and Peter Rogers, *Land Use and the Pipe* (Lexington, MA: Lexington Books, 1976).

41. James Heaney, Robert Pitt, and Richard Field, *Innovative Urban Wet-Weather Flow Management Systems* (Washington, DC: U.S. Environmental Protection Agency, National Risk Management Research Laboratory, Office of Research and Development, 1999).

42. James Wynn, *Innovative and Alternative On-Site Treatment of Residential Wastewater,* produced under U.S. Housing and Urban Development Grant RH-00-OH-I-0335, September 30, 2003, 21.

43. U.S. Environmental Protection Agency, Office of Water, *Wastewater Technology Fact Sheet: The Living Machine,* October 2002.

44. Stephen Dix and Valerie Nelson, "The Onsite Revolution: New Technology, Better Solutions," *Water Engineering & Management,* 145/10 (October 1998), 20.

45. See James Kreissl and Paul Chase, "Proposed National Onsite Standards: A Broad Assessment of Their Relative Benefits to Industry," *Small Flow Quarterly* 1.3/1 (winter 2002), 28–33.

46. The value of creating performance-based standards as a vehicle for implementing new concepts has also been acknowledged by government agencies such as the EPA. See, for example, the EPA's publication *Onsite Waste-Water Treatment Systems Manual* (Washington, DC: EPA, 2001).

47. Christine Rosen, *The Limits of Power: Great Fires and the Process of City Growth in America* (New York: Cambridge University Press, 1986).

48. David Wojick, "The Structure of Technological Revolutions," in George Bugliarello and Dean Doner, eds., *The History and Philosophy of Technology* (Urbana: University of Illinois Press, 1979).

Chapter 5

1. Fredrick Law Olmsted Jr., "Basic Principles of City Planning," in John Nolen, ed., *City Planning: A Series of Papers Presenting the Essential Elements of a City Plan* (New York: D. Appleton and Company, 1916), 1–18.

2. Sim Van der Ryn and Stuart Cowan, *Ecological Design* (Washington, DC: Island Press, 1996), 9.

3. Quoted in Stephen Seidel, *Housing Costs and Government Regulations: Confronting the Regulatory Maze* (New Brunswick, NJ: Center for Urban Policy Research, Rutgers, the State University of New Jersey, 1978), 119.

4. Jesse Clyde Nichols, "What We Have Learned," *Technical Bulletin No. 1* (Washington, DC: Urban Land Institute, 1945), 6.

5. Housing and Home Finance Agency, *Suggested Land Subdivision Regulations* (Washington, DC: U.S. Government Printing Office, 1952), 1.

6. American Society of Planning Officials, "Campaign Opened to 'Break' Municipal Subdivision Regulations and Control," *Planning Advisory Services News Letter,* November 23, 1945.

7. Charles Euchner, *Getting Home: Overcoming Barriers to Housing in Greater Boston* (Boston: Pioneer Institute for Public Policy Research and Rappaport Institute for Greater Boston, 2003), 42.

8. For the 1976 study see Stephen Seidel, *Housing Costs and Government Regulations: Confronting the Regulatory Maze* (New Brunswick, NJ: Center for Urban Policy Research, Rutgers, the State University of New Jersey, 1978). For the full study see Eran Ben-Joseph, *Subdivision Regulation: Practices and Attitudes—A Survey of Public Officials and Developers in the Nation's Fastest Growing Single Family Housing Markets,* WP03EB1 (Cambridge, MA: Lincoln Institute of Land Policy, July 2003).

9. Quoted in Ben-Joseph, *Subdivision Regulation,* 6–7.

10. In 1976, 72 percent of the developers surveyed indicated that unnecessary elements of subdivision regulations were responsible for significant inflation of the final selling price. In 2002, 59 percent of the developers surveyed stated that unnecessary elements of subdivision regulations were responsible for significant inflation of the final selling price.

11. Seidel, *Housing Costs and Government Regulations,* 125.

12. Seidel, *Housing Costs and Government Regulations,* 32.

13. Housing and Home Finance Agency, *Suggested Land Subdivision Regulations.*

14. In David Listokin and Carole Walker, *The Subdivision and Site Plan Handbook* (New Brunswick, NJ: Center for Urban Policy Research, 1989), 177.

15. In Michael Southworth and Eran Ben-Joseph, *Streets and the Shaping of Towns and Cities* (Washington, DC: Island Press, 2003).

16. In Dallas, for example, potential amounts of water not returned to the ground annually range from 6.2 to 14.4 billion gallons, while in Atlanta the amounts can reach 132.8 billion gallons, or enough water to supply the average daily household needs of 1.5 to 3.6 million

people per year. See American Rivers, Natural Resources Defense Council, and Smart Growth America, *Paving Our Way to Water Shortages: How Sprawl Aggravates the Effects of Drought* (Washington, DC: Smart Growth America, 2002).

17. Jonathon Levine and Aseem Inam, "Developer-Planner Interaction in Accessible Land Use Development," paper delivered at the 2001 Conference of the Association of Collegiate Schools of Planning, Cleveland, Ohio, November 2001.

18. Euchner, *Getting Home.*

19. U.S. Department of Housing and Urban Development, *Electronic Permitting Systems and How to Implement Them* (Washington, DC: U.S. Department of Housing and Urban Development, 2002), 20.

Chapter 6

1. Quoted in Sam Barnes, "Working in a Rice Paddy," *Louisiana Contractor,* November 2003 (http://louisiana.construction.com/features/archive/0311_feature2.asp).

2. American Rivers, Natural Resources Defense Council, and Smart Growth America, *Paving Our Way to Water Shortages: How Sprawl Aggravates the Effects of Drought* (Washington, DC: Smart Growth America, 2002).

3. Andy Reese, "Stormwater Paradigms," *Stormwater,* January 2004 (http://www.stormh2o .com/sw.html).

4. Gordon England, David Dee, and Stuart Stein, *Storm Water Retrofitting Techniques for Existing Development* (Viera, FL: Brevard County Surface Water Improvement, 2000).

5. See U.S. Department of Housing and Urban Development, *The Practice of Low Impact Development* (Washington, DC: U.S. Department of Housing and Urban Development, Office of Policy Development and Research, 2003), 31.

6. U.S. Department of Housing and Urban Development, *The Practice of Low Impact Development,* 31.

7. See the Pilot Project website at http://www.cityofseattle.net/util/SEAStreets/default.htm.

8. U.S. Environmental Protection Agency, *Guidance Specifying Management Measures for Sources of Nonpoint Pollution in Coastal Waters* (Washington, DC: USEPA, Office of Water, 1993).

9. Kathleen Corish, *Clearing and Grading Strategies for Urban Watersheds* (Washington, DC: Metropolitan Washington Council of Governments, 1995).

10. Quoted in Michael Michelsen, "Construction in the Space Age," *Grading and Excavation,* March-April 2000 (http://www.forester.net/gec.html).

11. Lawrence Freiser, ed., *California Government Tort Liability Practice* (Berkeley, CA: Continuing Education of the Bar, 1992), 367–372.

Chapter 7

1. The chapter epigraph is from Jesse Clyde Nichols, *What We Have Learned,* Technical Bulletin No. 1 (Washington, DC: Urban Land Institute, 1945), 6.

2. Community Associations Institute, *National Survey of Community Association Homeowner Satisfaction* (Alexandria, VA: Community Associations Institute Research Foundation, 1999).

3. Clifford Treese, *Community Association Factbook* (Alexandria, VA: Community Associations Institute, 1999).

4. Stephen Barton and Carol Silverman, eds., *Common Interest Communities: Private Governments and the Public Interest* (Berkeley, CA: Institute of Governmental Studies Press, 1994); Edward Blakely and Mary Snyder, *Fortress America* (Washington, DC: Brookings Institution Press, 1997); Robert Nelson, "Privatizing the Neighborhood: A Proposal to Replace Zoning with Private Collective Property Rights to Existing Neighborhoods," *George Mason Law Review* 827 (1999); Evan McKenzie, *Privatopia: Homeowner Associations and the Rise of Residential Private Government* (New Haven, CT: Yale University Press, 1994); Evan McKenzie, "Common-Interest Housing in the Communities of Tomorrow," *Policy Debate* 14 (2003), 203–234.

5. McKenzie, "Common-Interest Housing in the Communities of Tomorrow," 207.

6. Mike Davis, *City of Quartz* (London: Verso, 1990); Joel Garreau, *Edge City: Life on the New Frontier* (New York: Doubleday, 1991).

7. Blakely and Snyder, *Fortress America;* Robert Lang and Karen Danielsen, "Gated Communities in America: Walling the World Out," *Housing Policy Debate* 84 (1997), 867–899; Paula Franzese, "Does It Take a Village? Privatization, Patterns of Restrictiveness and the Demise of Community," *Villanova Law Review,* Villanova University, 47 (2002), 553.

8. Barton and Silverman, *Common Interest Communities.*

9. Peter Marcuse, "Walls of Fear and Walls of Support," in Nan Ellin, ed., *Architecture of Fear,* 101–114 (New York: Princeton Architectural Press, 1997).

10. Quoted in McKenzie, "Common-Interest Housing in the Communities of Tomorrow," 224.

11. Lang and Danielsen, "Gated Communities in America."

12. Community Associations Institute, *National Survey of Community Association Homeowner Satisfaction.*

13. Thomas Sanchez and Robert Lang, "Security versus Status: The Two Worlds of Gated Communities," *Metropolitan Institute at Virginia Tech Census Note* 02/02 (November 2002).

14. According to the authors, that may be partly due to the fact that there is a large Hispanic population in the West and Southwest, areas with the largest concentration of gated communities.

15. Community Associations Institute, *National Survey of Community Association Homeowner Satisfaction.*

16. *The Economist*, "Shut Up," November 30, 2002, 49.

17. Ulrich Jürgens and Martin Gnad, "Gated Communities in South Africa—Experiences from Johannesburg," *Environment and Planning B: Planning and Design* (2002), 337–353; Steven Robins, "At the Limits of Spatial Governmentality: A Message from the Tip of Africa," *Third World Quarterly—Journal of Emerging Areas* 23/4 (2002), 665–689.

18. Georg Glasze and Abdallah Alkhayyal, "Gated Housing Estates in the Arab World: Case Studies in Lebanon and Riyadh, Saudi Arabia," *Environment and Planning B: Planning and Design* 29 (2002), 321–336.

19. Pu Miao, "Deserted Streets in a Jammed Town: The Gated Community in Chinese Cities and Its Solution," *Journal of Urban Design* 8/1 (2003), 45–66.

20. Harald Leisch, "Gated Communities in Indonesia," *Cities* 19/5 (2002), 341–350.

21. Martin Coy and Martin Pöhler, "Gated Communities in Latin American Megacities: Case Studies in Brazil and Argentina," *Environment and Planning B: Planning and Design* 29/3 (May 2002), 355–370.

22. Pedro Pírez, "Buenos Aires: Fragmentation and Privatization of the Metropolitan City," *Environment and Urbanization* 14/1 (2002), 145–158.

23. Private developments were defined as developments that have private streets and are governed by HOA. For the full survey results and quotes see: Eran Ben-Joseph, *Subdivision Regulations: Practices and Attitudes—A Survey of Public Officials and Developers in the Nation's Fastest Growing Single-Family Housing Market,* WPO3EB1 (Cambridge, MA: Lincoln Institute of Land Policy, 2003).

24. Eran Ben-Joseph, *Residential Street Standards and Neighborhood Traffic Control: A Survey of Cities' Practices and Public Officials' Attitudes* (Berkeley: Institute of Transportation Studies, University of California at Berkeley, 1995).

25. U.S. Department of Housing and Urban Development, *The Practice of Low Impact Development* (Washington, DC: Office of Policy Development and Research, 2003).

26. Urban Land Institute, *Project Reference File* 24,15 (1994), 15.

27. Michael Southworth and Eran Ben-Joseph, *Streets and the Shaping of Towns and Cities* (Washington, DC: Island Press, 2003).

28. Dewees Island Property Owners Association, *Dewees Island Architectural and Environmental Design Guidelines* (Dewees Island, NC: Dewees Island Community, 1996), 1.

29. Dewees Island Property Owners Association, *Dewees Island Architectural and Environmental Design Guidelines,* 14.

30. David Takesuye, "Dewees Island: A Pioneering Example," *Urban Land* (Washington, DC: Urban Land Institute, May 2002), 28–29.

31. Jeff Rapson, "Private Wilderness Playgrounds: Understanding the Competitive Effects of Environmentally Oriented Master-Planned Communities" (master's thesis, Department of Urban Studies and Planning, MIT, 2002).

32. Rapson, *Private Wilderness Playgrounds.*

33. S. Seidel, *Housing Costs and Government Regulations: Confronting the Regulatory Maze* (New Brunswick, NJ: Center for Urban Policy Research, Rutgers, the State University of New Jersey, 1978).

34. McKenzie, "Common-Interest Housing in the Communities of Tomorrow," 203–234.

Chapter 8

1. Edgar Doctorow, *World's Fair* (New York: Plume, 1996), 252.

2. See, for example, Nancy Obermeyer, "Bureaucratic Factors in the Adaptation of GIS and Public Organizations: Preliminary Evidence from Public Administrators and Planners," *Computers, Environments and Urban Systems* 14 (1991), 261–271.

3. See Peter Bosselmann, *Representation of Places: Reality and Realism in City Design* (Berkeley: University of California Press, 1998); Peter Bosselmann, *Visual Simulation in Urban Design,* Working Paper 587 (Berkeley: Institute of Urban and Regional Development, University of California at Berkeley, 1992).

4. Donald Greenberg, "Computers and Architecture," *Scientific American,* February 1991, 104–109; Mark Weiser, "The Computer for the 21st Century," *Scientific American,* Sept. 1991, 94–104; Brad Myers, "A Brief History of Human Computer Interaction Technology," *ACM Interactions* 5/2 (March 1998), 44–54.

5. Peter Kamnitzer, "Computers and Urban Problems," in Harold Sackman and Harold Borko, eds., *Computers and the Problems of Society* (Reston, VA: AFIFP Press, 1972), 263–338.

6. Hiroshi Ishii and Brygg Ullmer, "Tangible Bits: Towards Seamless Interfaces between People, Bits and Atoms," in *Proceedings of CHI 97,* 22–27.

7. Weiser, "The Computer for the 21st Century."

8. See Pierre Wellner, "Interacting with Paper on the Digital Desk," *ACM* 36/7 (1993), 87–96; Wendy Mackay and Daniele Pagani, "Video Mosaic: Laying Out Time in a Physical Space," in *Proceedings of the Second ACM International Conference on Multimedia,* 1994, 165–172.

9. Ernesto Arias, Eden Hal, Gerhard Fischer, Andrew Gorman, and Eric Scharff, "Transcending the Individual Human Mind—Creating Shared Understanding through Collaborative Design," *ACM Transactions on Computer-Human Interaction (CHI)* 7/1, 2000, 84–113.

10. Morten Fjeld, Fred Voorhorst, Martin Bichsel, and Helmut Krueger, "Navigation Methods for an Augmented Reality System," *Abstracts of the CHI 2000* (The Hague: ACM Press, 2000), 8–9.

11. For the work of the Tangible Media Group, see Hiroshi Ishii and Tangible Media Group, *Tangible Bits: Towards Seamless Interface between People, Bits, and Atoms* (Tokyo: NTT Publishing, 2000). Also see their website at http://tangible.media.mit.edu/.

12. See John Underkoffler and Hiroshi Ishii, "Urp: A Luminous-Tangible Workbench for Urban Planning and Design," in *Proceedings of the CHI 99 Conference on Human Factors in Computing Systems: The CHI Is the Limit,* 386–393.

13. Ben Piper, Carlo Ratti, and Hiroshi Ishii, "Illuminating Clay: A 3-D Tangible Interface for Landscape Analysis," in *Proceedings of the CHI 02 Conference on Human Factors in Computing Systems,* 355–362.

14. Judith Innes, *Information in Communicative Planning,* Working Paper 679 (Berkeley: Institute of Urban and Regional Development, University of California at Berkeley, 1996).

15. Michael Batty, David Chapman, Steve Evans, Mordechai Haklay, Stefan Kueppers, Naru Shiode, Andy Smith, and Paul Torrens, "Visualizing the City: Communicating Urban Design to Planners and Decision Makers," in Richard Brail and Richard Klosterman, eds., *Planning Support Systems* (New Brunswick, NJ: ESRI Press and Center for Urban Policy Research, 2001), 406.

16. Case derived from http://www.asu.edu/caed/proceedings98/Chan/chan.html.

17. Michael Sable, "Ph.D. General Examination," unpublished manuscript, Department of Urban Studies and Planning, MIT, 12.

18. Case derived from http://www.asu.edu/caed/proceedings98/Chan/chan.html.

19. Tony Knight, "Virtual L.A. in the Works," San Jose, CA: *San Jose Mercury News,* August 11, 1997, 5E.

20. See Willa Reinhard, "Virtual Main Street: A New Computer Program Helps Communities Plan the Future," *Preservation Online, the online magazine of the National Trust for Historic Preservation,* September 20, 2002 (www.preservationonline.org).

21. Marcy Allen, CommunityViz, Boulder, CO, December 11, 2003.

Chapter 9

1. William Baer, "Toward Design of Regulations for the Built Environment," *Environment and Planning B: Planning and Design* 24 (1997), 37–57; David Rouse, Nancy Zobi, and Graciela Cavicchia, "Beyond Euclid: Integrating Zoning and Physical Planning," *Zoning News,* October 2001, 2–6.

2. Saad Yahya, Elijah Agevi, Lucky Lowe, Alex Mugova, Oscar Musandu-Nayamaryo, and Theo Schilderman, *Double Standards, Single Purpose: Reforming Housing Regulations to Reduce Poverty* (London: ITDG Publishing, 2001).

3. Anthony Downs, "Local Regulations and Housing Affordability," in Eran Ben-Joseph and Terry Szold, eds., *Regulating Place: Standards and the Shaping of Urban America* (New York: Routledge, 2005).

4. Although legislative bodies are not obligated to adopt a model code and may develop their own, studies conducted by the federal government have indicated that 97 percent of all U.S. cities with a building code have adopted the National Building Code. Three organizations—the Building Officials and Code Administrators (BOCA) International, Inc., International Conference of Building Officials (ICBO), and Southern Building Code Congress International, Inc. (SBCCI)—have also been created to handle the code enforcement.

5. International Code Council: International Code Adaptation (ICC 2003), at http://www.iccsafe.org/government/adoption.html.

6. Alexander Cuthbert, "Going Global: Reflexivity and Contextualism in Urban Design Education," *Journal of Urban Design* 6/3 (2001), 297–316.

7. Lawrence Vale, "Urban Design for Urban Development," in Lloyd Rodwin and Bishwapriya Sanyal, eds., *The Profession of City Planning: Changes, Images, and Challenges, 1950–2000* (New Brunswick, NJ: Center for Urban Policy Research, Rutgers, the State University of New Jersey, 2000), 220.

8. Michael Neuman, "Regional Design: Recovering a Great Landscape Architecture and Urban Planning Tradition," *Landscape and Urban Planning* 47 (2000), 115–128; Stephen Wheeler, "The New Regionalism: Key Characteristics of an Emerging Movement," *Journal of the American Planning Association* 68/3 (2002), 267–278.

9. Philip Angelides, "State Treasurer Phil Angelides Touts New Urbanism/Smart Growth," *The Planning Report* 8/7 (March 1999); Milt Hays, "The New Urbanism, Planning, and the

Failure of Political Imagination in Florida," *Bulletin of Science, Technology & Society* 20/4 (2000), 275–284; Dough Kelbaugh, "Three Paradigms: New Urbanism, Everyday Urbanism, Post Urbanism—An Excerpt from the Essential Common Place," *Bulletin of Science, Technology & Society* 20/4 (2000), 285–289; Chris Ellis, "The New Urbanism: Critiques and Rebuttals," *Journal of Urban Design* 7/3 (2002), 261–291.

10. Charles Bohl, *Place Making: Developing Town Centers, Main Streets, and Urban Villages* (Washington, DC: Urban Land Institute, 2002), 1.

11. Leigh Catesby, "Teaching Tradition," *The American Enterprise* 13/1 (2002), 36–41.

12. Dowell Myers, "The New Century's Boom in Planning School Enrollments," *Planetizen* Op-Ed, June 24, 2002 (www.Planetizen.org); Marice Chael, "New Urbanism at the University Level: Student New Urbanists Organize Chapters," *The Town Paper* 4/4 (2002), 2.

13. Alex Krieger, "The Planner as Urban Designer: Reforming Planning Education in the New Millennium," in Lloyd Rodwin and Bishwapriya Sanyal, eds., *The Profession of City Planning: Changes, Images, and Challenges, 1950–2000* (New Brunswick, NJ: Center for Urban Policy Research, Rutgers, the State University of New Jersey, 2000), 208.

14. William Baer, "Toward Design of Regulations for the Built Environment," *Environment and Planning B: Planning and Design* 24 (1997), 37–57 (quote on 40).

15. Alexander Garvin, *The American City: What Works, What Doesn't* (New York: McGraw-Hill, 1996), 391–392.

16. Bishwapriya Sanyal, "Planning's Three Challenges," in Lloyd Rodwin and Bishwapriya Sanyal, eds., *The Profession of City Planning: Changes, Images, and Challenges, 1950–2000* (New Brunswick, NJ: Center for Urban Policy Research, Rutgers, the State University of New Jersey, 2000).

17. Institute of Transportation Engineers, *Traditional Neighborhood Development Street Design Guidelines* (Washington, DC: ITE, 1999), 5.

18. Neil Peirce, "Zoning: Ready to Be Reformed?", *Washington Post Writers Group,* http://www.postwritersgroup.com/archives/peir0127.htm).

19. See http://web.mit.edu/sigus/www/.

20. For Christie Walk, see http://www.urbanecology.org.au/christiewalk/factsheet/. For Civano, see http://www.civano.com/.

21. See www.bestpractices.org.

22. See www.regbarriers.org.

23. See www.sustainable.doe.gov.

24. Roger Caves, *Land Use Planning: The Ballot Box Revolution* (Newbury Park, CA: Sage, 1992); Samuel Staley, "Ballot-Box Zoning, Transaction Costs, and Urban Growth," *Journal of the American Planning Association* 67/1 (1999), 25–37.

25. Keith Watson and Steven Gold, *The Other Side of Devolution: Shifting Relationships between State and Local Governments* (Washington, DC: Urban Institute, 1997), web:http://newfederalism.urban.org/html/other.htm.

26. U.S. Advisory Commission on Regulatory Barriers to Affordable Housing, *Not in My Back Yard: Removing Regulatory Barriers to Affordable Housing* (U.S. Government Printing Office, Washington, DC, 1991), 2–6.

27. Patricia Salkin, "Barriers to Affordable Housing: Are Land Use Controls the Scapegoat?", *Land Use Law & Zoning Digest* 45/4 (1993), 3–7; David Rouse, Nancy Zobi, and Graciela Cavicchia, "Beyond Euclid: Integrating Zoning and Physical Planning," *Zoning News,* October 2001; American Planning Association, *Growing Smart: Legislative Guidebook* (Chicago: American Planning Association, 2001), 8.

28. American Planning Association, *Growing Smart Legislative Guidebook,* 8–85.

29. Oregon Department of Transportation, *Neighborhood Street Design Guidelines: An Oregon Guide for Reducing Street Widths* (Salem, OR: Department of Transportation and the Department of Land Conservation and Development, 2000), 5.

30. The transect as a basis for coding has been used by the firms of Moule & Polyzoides, Dover/Kohl, Torti-Gallas, Mouzon & Associates, and Duany Plater-Zyberk & Company.

31. U.S. Department of Housing and Urban Development, *Electronic Permitting Systems and How to Implement Them* (Washington, DC: HUD, 2002).

32. M. Dennison, "Zoning and Comprehensive Plan," *Zoning News,* August 1996.

33. Garvin, *The American City,* American Planning Association, *Growing Smart: Legislative Guidebook.*

34. For examples of precise plans see the city of Mountain View, California, Advance Planning Devision at http://www.ci.mtnview.ca.us/citydepts/cd/apd/adv_plan.htm, and their various Precise Plans at http://www.ci.mtnview.ca.us/citydepts/cd/apd/precise_plans.htm.

35. For examples see http://www.vicgroup.com/ and http://urban-advantage.com.

Afterword

1. Bill Osinski, "Habersham Takes Ax to Government," *Atlanta Journal-Constitution,* December 23, 2002, B1.

2. Alexis de Tocqueville, *Democracy in America* (New York: Doubleday, 1966), 255.

Other References

Abercrombie, Patrick. *Town and Country Planning.* London: Oxford University Press, 1943.

Adams, Thomas. *Outline of Town and City Planning: A Review of Past Efforts and Modern Aims.* New York: Russell Sage Foundation, 1935.

AlSayyad, Nezar, ed. *Forms of Dominance on the Architecture and Urbanism of the Colonial Enterprise.* Brookfield, USA: Avebury, 1992.

American Society of Planning Officials. *ASPO Index to Proceedings of National Planning Conferences, 1909–1961.* Chicago: American Society of Planning Officials, 1962.

American Society of Planning Officials. *Problems of Zoning and Land-Use Regulation.* Washington, DC: American Society of Planning Officials, 1968.

Anderson, Robert. *Planning, Zoning & Subdivision: A Summary of Statutory Law in the 50 States.* New York: New York State Federation of Official Planning Organizations, 1966.

Angel, Shlomo. *Housing Policy Matters: A Global Analysis.* New York: Oxford University Press, 2000.

Antoniou, Jim. *Islamic Cities and Conservation.* Paris: UNESCO Press, 1981.

Argan, Giulio. *The Renaissance City.* New York: Braziller, 1970.

Aronovici, Carol. *Housing the Masses.* New York: Wiley, 1939.

Bair, Frederick. *The Text of a Model Zoning Ordinance, with Commentary.* Chicago: American Society of Planning Officials, 1966.

Banerjee, Tridib, and William Baer. *Beyond the Neighborhood Unit.* New York: Plenum Press, 1984.

Bardach, Eugene, and Robert A. Kagan. *Going by the Book: The Problem of Regulatory Unreasonableness.* Philadelphia: Temple University Press, 1982.

Beder, Sharon. *The New Engineer: Management and Professional Responsibility in a Changing World.* Melbourne, Australia: Macmillan, 1998.

Benevolo, Leonardo. *The Origins of Modern Town Planning*. Cambridge, MA: MIT Press, 1967.

Black, Russell. *Building Lines and Reservations for Future Streets: Their Establishment and Protection*. Cambridge, MA: Harvard University Press, 1935.

Black, Russell. *Planning the Small American City*. Chicago: Public Administration Service, 1944.

Blau, Judith, La Gory, Mark, and John Pipkin, eds. *Professionals and Urban Form*. Albany: State University of New York Press, 1983.

Bohl, Charles. *Place Making Developing Town Centers, Main Streets, and Urban Villages*. Washington, DC: Urban Land Institute, 2002.

Boyer, Christine. *The City of Collective Memory: Its Historical Imagery and Architectural Entertainments*. Cambridge, MA: MIT Press, 1994.

Boyer, Christine. *Dreaming the Rational City: The Myth of American City Planning*. Cambridge, MA: MIT Press, 1983.

Broadbent, Geoffrey. *Emerging Concepts in Urban Space Design*. London: Van Nostrand Reinhold (International), 1990.

Calhoun, Daniel. *The American Civil Engineer: Origins and Conflict*. Cambridge, MA: Technology Press, MIT; distributed by Harvard University Press, 1960.

Castells, Manual. *The Urban Question: A Marxist Approach*. Cambridge, MA: MIT Press, 1977.

Choay, Françoise. *La règle et le modèle: Sur la théorie de l'architecture et de l'urbanisme* (The Rules and Models: Theories of Architecture and Urbanism). Paris: Éditions du Seuil, 1980.

Churchill, Henry, and Roslyn Ittleson. *Neighborhood Design and Control: An Analysis of the Problem of Planned Subdivisions*. New York: National Committee on Housing, 1944.

Collins, Roseborough. *Camillo Sitte and the Birth of Modern City Planning*. New York: Random House, 1965.

Creese, Walter. *The Search for Environment: The Garden City, Before and After*. New Haven, CT: Yale University Press, 1966.

Cullen, Gordon. *The Concise Townscape*. New York: Van Nostrand Reinhold, 1971.

Cullingworth, John. *Planning in the USA: Policies, Issues, and Processes*. London: Routledge, 1997.

Dahir, James. *The Neighborhood Unit Plan, Its Spread and Acceptance: A Selected Bibliography with Interpretative Comments*. New York: Russell Sage Foundation, 1947.

Davis, Mike. *City of Quartz: Excavating the Future in Los Angeles*. New York: Verso, 1990.

De Chiara, Joseph. *Urban Planning and Design Criteria.* New York: Van Nostrand Reinhold, 1975.

De Chiara, Joseph, Julius Panero, and Martin Zelnik, eds. *Time-Saver Standards for Housing and Residential Development.* New York: McGraw-Hill, 1995.

Duany, Andres. *The New Civic Art: Elements of Town Planning.* New York: Rizzoli, 2003.

Duany, Andres. *Towns and Town-Making Principles.* New York: Rizzoli, 1991.

Duany, Andres, and Elizabeth Plater-Zyberk. *Towns and Town-Making Principles.* Ed. Alex Krieger. New York: Rizzoli, 1991.

Fishman, Robert. *Bourgeois Utopias: The Rise and Fall of Suburbia.* New York: Basic Books, 1987.

Foldvary, Fred. *Public Goods and Private Communities: The Market Provision of Social Services.* Brookfield, VT: Edward Elgar, 1994.

Ford, George. *Building Height, Bulk, and Form.* Cambridge, MA: Harvard University Press, 1931.

Fukuyama, Francis. *The Great Disruption: Human Nature and the Reconstitution of Social Order.* New York: Free Press, 1999.

Gallion, Arthur. *The Urban Pattern: City Planning and Design.* New York: Van Nostrand, 1975.

Geddes, Patrick. *Cities in Evolution.* London: Williams & Norgate, 1949.

Gibberd, Frederick. *Town Design.* London: Architectural Press, 1953.

Goodman, Robert. *After the Planners.* New York: Simon and Schuster, 1971.

Haverfield, Francis. *Ancient Town-Planning.* Oxford: Clarendon Press, 1913.

Hegemann, Werner. *The American Vitruvius: An Architect's Handbook of Civic Art.* New York: Architectural Book Publishing Co., 1922.

Jellicoe, Geoffrey. *Motopia: A Study in the Evolution of Urban Landscape.* New York: Praeger, 1961.

Kaiser, Edward. *Urban Land Use Planning.* Urbana: University of Illinois Press, 1995.

Kending, Lane. *Performance Zoning.* Chicago: American Planning Association, 1980.

Kostka, Joseph. *Planning Residential Subdivisions.* Winnipeg, Canada: Hignell Printing, 1954.

Kostof, Spiro. *The City Assembled: The Elements of Urban Form through History.* Boston: Little, Brown, 1992.

Kostof, Spiro. *The City Shaped: Urban Patterns and Meanings through History.* Boston: Bulfinch Press, 1991.

Krier, Leon. *Urban Transformations.* London: Architectural Design, AD Editions, 1978.

Krier, Rob. *Urban Space—Stadtraum.* New York: Rizzoli International Publications, 1979.

Lai, Richard. *Law in Urban Design and Planning: The Invisible Web.* New York: Van Nostrand Reinhold, 1988.

Lavedan, Pierre. *Histoire de l'urbanisme.* Paris: H. Laurens, 1959.

LeGates, Richard, and Frederic Stout, eds. *Early Urban Planning.* London: Routledge/Thoemmes Press, 1998.

Lewis, Harold. *Planning the Modern City.* New York: Wiley, 1949.

Listokin, David, and Carole Walker, eds. *The Subdivision and Site Plan Handbook.* New Brunswick, NJ: Center for Urban Policy Research, 1989.

Loyer, François. *Architecture of the Industrial Age, 1789–1914.* New York: Skira/Rizzoli, 1983.

Lynch, Kevin. *Site Planning.* Cambridge, MA: MIT Press, 1971.

Moore, Clement Henry. *Images of Development: Egyptian Engineers in Search of Industry.* Cairo: American University Press, 1994.

Moudon, Anne Vernez. *Built for Change: Neighborhood Architecture in San Francisco.* Cambridge, MA: MIT Press, 1986.

Mumford, Lewis. *City Development.* New York: Harcourt, Brace, 1945.

Mumford, Lewis. *The City in History: Its Origins, Its Transformations, and Its Prospects.* New York: Harcourt, Brace Jovanovich, 1961.

Mumford, Lewis. *The Culture of Cities.* New York: Harcourt, Brace Jovanovich, 1938.

Mumford, Lewis. The Neighborhood and the Neighborhood Unit. *Town Planning Review* 24/4 (January 1954), 258–262.

New York Regional Plan Association. *Regional Plan of New York and its Environs.* New York: NYRPA, 1929.

Osborn, Frederic. *Green Belt Cities.* London: Evelyn, Adams & Mackay, 1969.

Osborn, Frederic. *New Towns After the War.* London: Dent, 1942.

Paquette, Randor, Norman Ashford, and Paul Wright. *Transportation Engineering: Planning and Design.* New York: Ronald Press Co., 1972.

Perin, Constance. *Everything in Its Place.* Princeton, NJ: Princeton University Press, 1977.

Perin, Constance. *With Man in Mind.* Cambridge, MA: MIT Press, 1970.

Perry, Clarence. *Housing for the Machine Age.* New York: Russell Sage Foundation, 1939.

Ponting, Clive. *A Green History of the World: The Environment and the Collapse of Great Civilizations.* New York: Penguin Books, 1993.

Punter, John. *Design Guidelines in American Cities: A Review of Design Policies and Guidance in Five West Coast Cities.* Liverpool: Liverpool University Press, 1999.

Rapoport, Amos. *Human Aspects of Urban Form: Towards a Man-Environment Approach to Form and Design.* Oxford: Pergamon Press, 1977.

Rapoport, Amos. *The Meaning of the Built Environment: A Nonverbal Communication Approach.* Beverly Hills, CA: Sage, 1982.

Reps, John. *The Making of Urban America: A History of City Planning in the United States.* Princeton, NJ: Princeton University Press, 1965.

Ritter, Paul. *Planning for Man and Motor.* New York: Pergamon Press, 1964.

Robinson, Charles Mulford. *The Improvement of Towns and Cities; or, The Practical Basis of Civic Aesthetics.* New York: Putnam, 1901.

Robinson, Charles Mulford. *Modern Civic Art; or, The City Made Beautiful.* New York: Putnam, 1903.

Robinson, Charles Mulford. *The Width and Arrangement of Streets: A Study in Town Planning.* New York: The Engineering News Publishing Company, 1911.

Rowe, Peter. *Making a Middle Landscape.* Cambridge, MA: MIT Press, 1991.

Saarinen, Eliel. *The City, Its Growth, Its Decay, Its Future.* New York: Reinhold Publishing Corporation, 1943.

Sears, Roebuck and Company, Community Planning Division. *ABC's of Community Planning.* Chicago: Sears, Roebuck, 1962.

Sert, José Luis. *Can Our Cities Survive? An ABC of Urban Problems, Their Analysis, Their Solutions.* Cambridge, MA: Harvard University Press, 1942.

Sitte, Camillo. *The Art of Building Cities: City Building According to its Artistic Fundamentals.* New York: Reinhold Publishing Corporation, 1945.

Solomon, Daniel. *ReBuilding.* Princeton, NJ: Princeton Architectural Press, 1992.

Stein, Clarence. *Toward New Towns for America.* Cambridge, MA: MIT Press, 1951.

Steinhardt, Nancy. *Chinese Imperial City Planning.* Honolulu: University of Hawaii Press, 1990.

Stilgoe, John. *Borderland: Origins of the American Suburb 1820–1939*. New Haven, CT: Yale University Press, 1988.

Sucher, David. *City Comforts: How to Build an Urban Village*. Seattle: City Comforts Press, 1995.

Triggs, Inigo. *Town Planning, Past, Present and Possible*. London: Methuen & Co., 1909.

Unwin, Raymond. *Nothing Gained in Overcrowding*. London: Garden Cities and Town Planning Association, 1912.

Unwin, Raymond. *Town Planning in Practice: An Introduction to the Art of Designing Cities and Suburbs*. London: Ernest Benn Limited, 1911.

Urban Land Institute. *The Community Builders Handbook*. Washington, DC: ULI, 1947.

Warner, Kee, and Harvey Molotch. *Building Rules: How Local Controls Shape Community Environments and Economics*. Boulder, CO: Westview Press, 2000.

Warner, Sam Bass. *Streetcar Suburbs: The Process of Growth in Boston, 1870–1900*. Cambridge, MA: Harvard University Press, 1978.

Warner, Sam Bass. *The Urban Wilderness: A History of the American City*. New York: Harper & Row, 1972.

Weiss, Marc. *The Rise of the Community Builder*. New York: Columbia University Press, 1987.

Whyte, William. *Cluster Development*. New York: American Conservation Association, 1964.

Williams, Frank. *The Law of City Planning and Zoning*. New York: Macmillan, 1922.

Wright, Henry. *Rehousing Urban America*. New York: Columbia University Press, 1935.

Index

Page numbers for illustrations are printed in boldface.